Network-on-Chip Architectures

AF789209

Lecture Notes in Electrical Engineering

Volume 45

For other titles published in the series, go to
www.springer.com/series/7818

Chrysostomos Nicopoulos
Vijaykrishnan Narayanan • Chita R. Das

Network-on-Chip Architectures

A Holistic Design Exploration

Springer

Chrysostomos Nicopoulos
Department Electrical and Computer Engineering
University of Cyprus
Cyprus
nicopoulos@ucy.ac.cy

Vijaykrishnan Narayanan
Department Computer Science and Engineering
Pennsylvania State University
University Park PA
USA

Chita R. Das
Department of Computer Science and Engineering
Pennsylvania State University
University Park PA
USA

ISBN 978-94-007-3049-6 e-ISBN 978-90-481-3031-3
DOI 10.1007/978-90-481-3031-3
Springer Dordrecht Heidelberg London New York

© Springer Science+Business Media B.V. 2010
Softcover reprint of the hardcover 1st edition 2010
No part of this work may be reproduced, stored in a retrieval system, or transmitted in any form or by any means, electronic, mechanical, photocopying, microfilming, recording or otherwise, without written permission from the Publisher, with the exception of any material supplied specifically for the purpose of being entered and executed on a computer system, for exclusive use by the purchaser of the work.

Printed on acid-free paper

Springer is part of Springer Science+Business Media (www.springer.com)

Abstract

The continuing reduction of feature sizes into the nanoscale regime has led to dramatic increases in transistor densities. Computer architects are actively pursuing multi-core designs with billions of transistors on a single die. Integration at these levels has highlighted the criticality of the on-chip interconnects; global interconnect delays are dominating gate delays and affecting overall system performance. Packet-based Network-on-Chip (NoC) architectures are viewed as a possible solution to burgeoning global wiring delays in many-core chips, and have recently crystallized into a significant research domain. NoCs are steadily becoming the de facto interconnect solution in complex Systems-on-Chip (SoC), because of their scalability and optimized electrical properties. However, current research also indicates that the chip area and power budgets are increasingly being dominated by the interconnection network. To combat this escalating trend, attention should be paid to the optimization of the interconnect architecture.

Unlike traditional multi-computer macro-networks, on-chip networks instill a new flavor to communication research due to their inherently resource-constrained nature. Scarcity in the area and power budgets devoted to the interconnection fabric necessitates a re-interpretation of the networking paradigm. Furthermore, despite the lightweight character demanded of the NoC components, modern designs require ultra-low communication latencies in order to cope with inflating data bandwidths. These conflicting requirements transform the NoC design process into a grand challenge for the system designer. The work presented in this volume aims to address these issues through a comprehensive and holistic exploration of the design space. To truly appreciate the nuances underlying the NoC realm, the design aspects of the on-chip network are viewed through a penta-faceted prism encompassing five major issues: (1) ***performance***, (2) ***silicon area consumption***, (3) ***power/energy efficiency***, (4) ***reliability***, and (5) ***variability***. These five aspects serve as the fundamental design drivers and critical evaluation metrics in the quest for efficient NoC implementations.

The research described in this volume explores the field by employing a two-pronged approach: (a) ***MICRO-architectural innovations*** within the major NoC components, and (b) ***MACRO-architectural choices*** aiming to seamlessly merge the interconnection backbone with the remaining system modules. These two research threads, along with the aforementioned five key metrics mount a holistic and in-depth attack on most issues surrounding the design and integration of NoCs in modern multi-core architectures. Based on this premise of two complementary core themes, the volume is divided into two corresponding parts; the first part delves into the world of MICRO-architectural exploration of the NoC paradigm, while the second part shifts the focus to a MACRO-architectural abstraction level. Ultimately, both parts work in unison in attacking several pressing issues concerning on-chip interconnects in the new multi/many-core reality.

Acknowledgments

My immense appreciation and wholehearted gratitude goes to my thesis advisor, Professor Vijaykrishnan Narayanan. I am deeply indebted to him for his constant encouragement, unending support, insightful guidance, and wisdom in all aspects of life. Being a mentor and role model, he exposed me to countless unique opportunities in the field of my studies, and gave me the unique opportunity to interact extensively with world-class researchers in the area of computer engineering. His inexhaustible enthusiasm and inspirational motivation have helped me grow both as a student and as a person. With him at the helm, I have enjoyed the most exhilarating, enjoyable, and rewarding ride of my life; and for that, I will always be eternally grateful. Most importantly, beyond the thrills of a myriad research endeavors, he cared – above all – about my happiness and success in life.

I would also like to thank Professor Ken Jenkins for serving as my thesis co-advisor and for his insightful comments and suggestions.

I am sincerely grateful to Professor Chita R. Das for his invaluable contributions to my work over the last few years. His vast wealth of expertise, broad vision, and perceptive wisdom have led to a hugely successful and fruitful collaboration. It is this collaboration that forms the foundation of this volume.

I am equally indebted to Professor Mary Jane Irwin for her steadfast support and precious advice throughout my presence in the Microsystems Design Lab. I am especially grateful to her for exposing me to a multitude of opportunities and giving me the chance to gain enormous experience and blossom as a researcher.

I am extremely grateful to Professors Yuan Xie and Mahmut Kandemir for their sincere friendship, enlightening discussions, priceless insights, and intellectual contributions during my studies. Furthermore, I would like to thank Professor Ram Narayanan for graciously serving on my Ph.D. thesis committee and for his perceptive and encouraging comments. I also wish to thank Professor Paul S. Weiss for instilling a scientific flavor in the committee and – as the external member – bringing a different, yet invaluable perspective to the table.

Eternal gratitude and my sincerest appreciation go to my colleagues and dear friends Jongman Kim and Dongkook Park for a very productive, supportive, successful, and enjoyable collaboration. I would also like to thank Thomas D. Richardson and Feihui Li for being great friends and collaborators; their material and intellectual contributions to my work are priceless. Many thanks go to Jungsub Kim, Suresh Srinivasan, Aditya Yanamandra, Reetuparna Das, Feng Wang, and Theocharis

Theocharides for their enormous help and support. I am also profoundly grateful to all past and present members of the Microsystems Design Lab for the amazing and unforgettable memories of the last few years.

Many thanks also go to all my friends and relatives for their unremitting and resolute confidence in me and their constant encouragement and support.

Finally, I owe my deepest gratitude and boundless appreciation to my parents, Andreas and Elpiniki, for their endless love, unwavering support, constant encouragement, and unfaltering and sincere belief in me. Without them, this volume would not have happened…

Contents

List of Figures

List of Tables

List of Abbreviations

AC	Allocation comparator
ACK	Acknowledgment
ACLV	Across chip line-width variations
AD	Adaptive
AG	Affinity group
ASIC	Application specific integrated circuits
BC	Bit-complement
BIST	Built-in self-test
CB	Connection box
CC	Cache compression
CDMA	Code-division multiple access
CMOS	Complementary metal-oxide semiconductor
CMP	Chip multi-processor
CPU	Central processing unit
DAMQ	Dynamically allocated multi-queue
DAMQWR	DAMQ-with-recruit-registers
DEMUX	De-multiplexer (i.e. decoder)
DimDe	Dimensionally decomposed router
DLT	Direction lookup table
DN	Direct neighbor
DNUCA	Dynamic non-uniform cache architecture
DOR	Dimension-order routing
DRAM	Dynamic random access memory
DT	Deterministic
dTDMA	Dynamic time-division multiple access
E2E	End-to-end
ECC	Error correcting codes
EDP	Energy-delay product
EM	Electromigration
F2B	Face-to-back
F2F	Face-to-face
FBF	Fullest buffer first

FC-CB	Fully connected circular buffer
FCFS	First-come first-served
FEC	Forward error correction
FIFO	First-in first-out
FP	Fast path
HBH	Hop-by-hop
HCE	Hot carrier effects
HDL	Hardware description language
HoL	Head-of-line
IIL	Inter-intra layer
ILP	Instruction-level parallelism
IN	Indirect neighbor
IP	Intellectual property
IS	Information set
LCR	Leakage classification register
LWF	Longest wait first
MEMS	Micro-electro-mechanical systems
MEP	Multiple entry point
MESI	Modified-exclusive-shared-invalid
MLBS	Multi-layer buried structures
MPSoC	Multi-processor system-on-chip
MSI	Modified-shared-invalid
MUX	Multiplexer
NACK	Negative acknowledgment
NBTI	Negative bias temperature instability
NC	Network interface controller compression
NetInMem	Network-in-memory
NIC	Network interface controller
NoC	Network-on-chip
NR	Normal random
NUCA	Non-uniform cache architecture
OPC	Optical proximity correction
PA	Proximity-aware
PBD	Platform-based design
PC	Physical channel
PDP	Power-delay product
PE	Processing element
PEF	Performance, energy and fault-tolerance
PFA	Path frequency analyzer
PFT	Path frequency table
PIM	Parallel iterative matching

PS	Path set
PTM	Predictive technology model
PV	Process variation
QoS	Quality-of-service
RC	Routing computation
RDF	Random dopant fluctuations
RoCo	Row-column decoupled router
RT	Routing table
RTL	Register-transfer level
SA	Switch allocation
SEC/DED	Single error correction and double error detection
SER	Soft-error rate
SIMD	Single-instruction multiple-data
SIP	Service information provider
SNUCA	Static non-uniform cache architecture
SoC	System-on-chip
SOI	Silicon-on-insulator
SON	Service-oriented networking
SS	Self-similar
ST	Sliding timeslice
STB	Set-top box
STI	Sony, Toshiba, and IBM
TDDB	Time-dependant dielectric breakdown
TDMA	Time-division multiple access
TLP	Thread-level parallelism
TMR	Triple module redundancy
TN	Tornado
TPA	Thermal proximity-aware
TSMC	Taiwan Semiconductor Manufacturing Company
TTM	Time-to-market
UBS	Unified buffer structure
UCL	Unified control logic
UR	Uniform random
UTP	Unique-Token protocol
VA	Virtual channel arbitration
VC	Virtual channel
VCDAMQ	Virtual channel dynamically allocated multi-queue
ViChaR	Virtual channel regulator
XBAR	Crossbar

Chapter 1
Introduction

Since the beginning of the new millennium, the world of digital system design has witnessed an unprecedented phenomenon: a rapid and persistent reduction in feature sizes well into the nanoscale realm. Advancements in device technology and fabrication techniques have enabled designers to tread into previously unchartered territories; integration of billions of transistors on-die is now a reality. At such integration levels, it is imperative to employ parallelism to effectively utilize the transistors [1].

1.1 The Diminishing Returns of Instruction-Level Parallelism

In order to extract as much parallelism out of the processing engine as possible, today's microprocessors incorporate a multitude of sophisticated micro-architectural features, such as multiple instruction issue, dynamic scheduling, out-of-order execution, speculative execution and dynamic branch prediction. However, in order to sustain performance growth, future superscalar microprocessors must rely on even more complex architectural innovations. Olukotun et al. [2] have showed that circuit limitations and limited instruction level parallelism will diminish the benefits afforded to the superscalar model by increased architectural complexity. Increased issue widths cause a quadratic increase in the size of issue queues and the complexity of register files. Furthermore, as the number of execution units increases, wiring and interconnection logic complexity begin to adversely affect performance. This newly-developed ability to integrate previously unimaginable amounts of logic on a single die has naturally led to a paradigm shift in computer architecture. Instead of concentrating on exploiting the saturating Instruction-Level Parallelism (ILP) through complex, super-scalar cores, architects are now starting to target Thread-Level Parallelism (TLP) by using multiple simple Central Processing Units (CPU) cores. The embodiment of this transition is the advent of Chip Multiprocessors (CMP) as a viable alternative to the complex superscalar architecture. CMPs are simple, compact processing cores forming a decentralized micro-architecture which scales more efficiently with increased integration densities

C. Nicopoulos et al., *Network-on-Chip Architectures: A Holistic Design Exploration*, Lecture Notes in Electrical Engineering 45,
DOI 10.1007/978-90-481-3031-3_1, © Springer Science+Business Media B.V. 2009

[2]. The Cell Processor from Sony, Toshiba and IBM (STI) [3], and the Sun UltraSPARC T1 (formerly codenamed Niagara) [4] signal the growing popularity of such systems. Furthermore, Intel's very recently announced 80-core TeraFLOP chip [5] exemplifies the irreversible march toward many-core systems with tens or even hundreds of processing elements.

1.2 The Dawn of the Communication-Centric Revolution

The multi-core thrust has ushered the gradual displacement of the computation-centric design model by a more communication-centric approach [6]. The large, sophisticated monolithic modules are giving way to several smaller, simpler processing elements working in tandem. This trend has led to a surge in the popularity of multi-core systems, which typically manifest themselves in two distinct incarnations: heterogeneous Multi-Processor Systems-on-Chip (MPSoC) and homogeneous Chip Multi-Processors (CMP). The SoC philosophy revolves around the technique of Platform-Based Design (PBD) [7], which advocates the reuse of Intellectual Property (IP) cores in flexible design templates that can be customized accordingly to satisfy the demands of particular implementations. The appeal of such a modular approach lies in the substantially reduced Time-To-Market (TTM) incubation period, which is a direct outcome of lower circuit complexity and reduced design effort. The whole system can now be viewed as a diverse collection of pre-existing IP components integrated on a single die. Naturally, the correctness and efficiency of operation rely heavily on inter-module communication. In fact, this new design paradigm has inflicted enormous strain on the interconnection backbone, which now needs to undertake a more prominent and sophisticated role.

1.3 The Global Wiring Challenge

The surge in popularity of SoC architectures has, therefore, highlighted the importance of the on-chip interconnection fabric and has drawn attention to the intricacies surrounding the wiring aspects of complex chip design.

Modern Application Specific Integrated Circuits (ASIC) consist of a number of metal layers (around eleven, as of 2006 [8]) situated on top of the active silicon layer. A typical layout is illustrated in Fig. 1.1 [8]. These layers are used to interconnect the various components of the chip. Bottom metal layers tend to be thin and narrow, while top layers are thicker and wider. As shown in Fig. 1.1, metal layers are divided into four distinct groups, based on the length of the respective interconnects: (1) Metal 1 (also referred to as Local), (2) Intermediate, (3) Semi-Global, and (4) Global. The top metal layers (i.e., Global) are used – as their name suggests – for

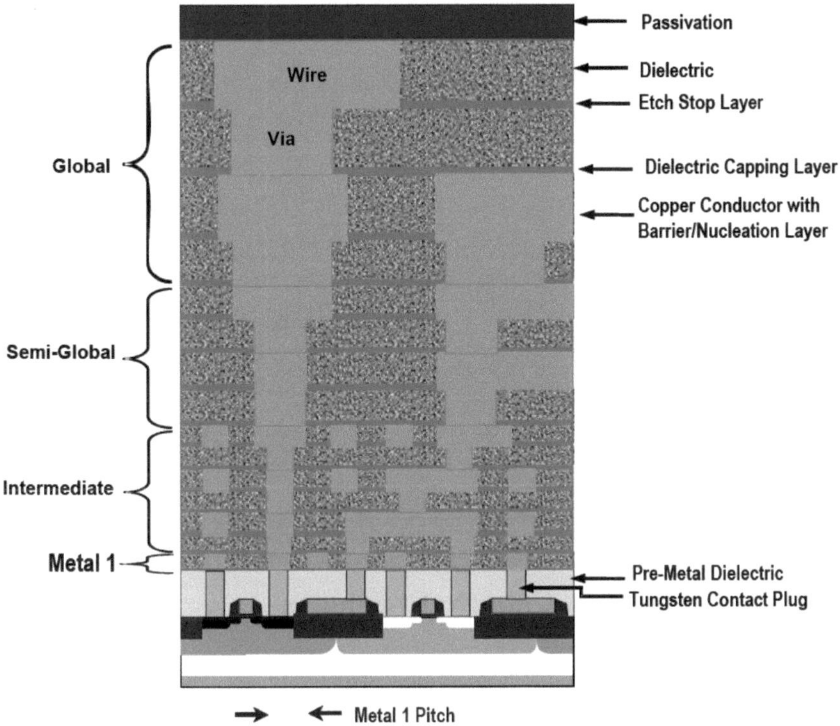

Fig. 1.1 Metal layers in modern ASIC devices [8]

long interconnects between distant components on the chip. The larger cross-section of these wires (a corollary of their larger width and thickness) ensures lower resistance and, therefore, increased propagation speed.

While gate (i.e., logic) delays have been scaling down dramatically over the last few years, global wiring delays are, instead, increasing [8]; as wire cross-sections decrease, resistance increases. This troublesome development is illustrated in Fig. 1.2 [8], which plots relative delay against process technology size. The top two curves on the graph highlight the alarming increase in global wiring delays, relative to gate delays, as technology feature sizes diminish into the nanoscale regime. Global wire delays increase exponentially with feature size, or at best linearly after repeater insertion [9]. Hence, this conundrum is transforming the interconnect backbone into the major performance bottleneck in modern, many-core systems, which require communication over substantial distances across the chip. The long communication distances are further compounded by increasing clock frequencies; the latter imply that signal traversal across the chip die will require many clock cycles [10], adversely affecting overall system performance. Alarmingly, it has been projected that in forthcoming technology nodes, up to 77% of the delay will be attributed to the interconnect [11].

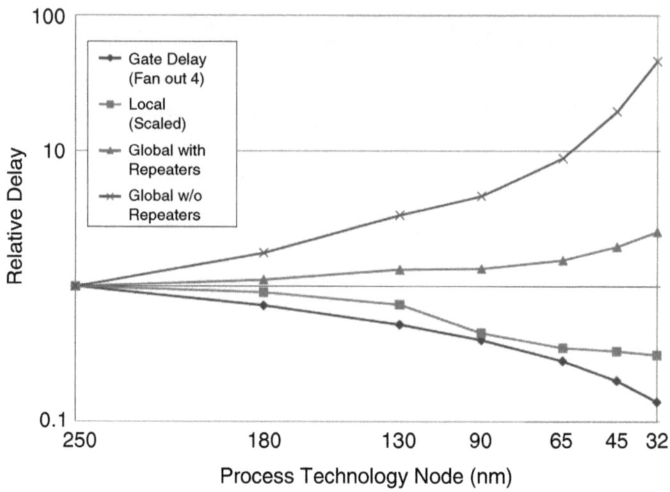

Fig. 1.2 Relative wire delays in ASIC implementations [8]

1.4 The Network-on-Chip (NoC) Solution

A variety of interconnection schemes are currently in use, including crossbars, rings, buses, and NoCs [12]. Of these, the latter two have been dominant in the research community [13,14]. However, buses suffer from poor scalability; as the number of processing elements increases, performance degrades dramatically. Hence, they are not considered appropriate for systems of more than about 10 nodes [13,15]. To overcome this limitation, attention has shifted to packet-based on-chip communication networks, known as Networks-on-Chip (NoC) [13,14,16,17]. Much like their macro-network brethrens, NoCs scale very efficiently as the number of nodes (e.g., processing elements) increases. Due to their scalability, NoCs are considered the most viable solution for many-core chips of the future. Intel's TeraFLOP chip is a prime example of this assertion [5].

NoCs, however, pose several design challenges emanating from their inherently stringent resource constraints; namely, area and power limitations. Recent research results have indicated some worrisome trends pertaining to the NoC's consumption of overall system resources; the chip area and power budgets are increasingly being dominated by the interconnection network [6,18,19]. As the architectural focus shifts from centralized, monolithic designs to distributed, multi-core systems, communication power has become comparable to logic and memory power, and is expected to eventually surpass them [6]. In fact, on-chip communication already consumes a significant portion of the chip's power budget (about 40%) [20,21]. This ominous trend has been observed by several researchers [13,18,22,23] and the realization of its ramifications has fueled momentum in investigating NoC architectures. The on-chip network's significant consumption of valuable resources is a stark reminder of its criticality in digital system design.

Thus, all three critical objective functions in any chip design (namely area, power and performance) are ominously being dominated by the underlying interconnect. This realization is very significant, since its impact could shake the foundations of all future many-core designs. Ideally, the interconnect should not stand out as the prime consumer of on-chip resources. Instead, it should blend as seamlessly as possible within the system, and contribute its services without being the sticking thorn in the design space.

Furthermore, aggressive technology scaling has accentuated the issue of reliability due to a rapid increase in the prominence of permanent faults; these are mostly caused from accelerated aging effects such as Time Dependant Dielectric Breakdown (TDDB) [24–26], Electromigration (EM) [27], Negative Bias Temperature Instability (NBTI) and Hot Carrier Effects (HCE) [28–30]. Moreover, soft upsets caused by cross-talk, coupling noise, power supply noise, and transient faults are also a concern to overall reliability [31–33]. Finally, thermal stressing due to hot spots has a direct bearing on both performance and power consumption and brings another complexity to the design space. In particular, this will play an important role in future 3D chip architectures, where elevated power densities are fairly common [34]. The growing concern about reliability in the interconnection network has prompted extensive research in this area [35–37].

A new serious impediment is also emerging as a force to be reckoned with: Process Variation (PV) resulting from manufacturing imperfections. PV is observed due to random effects, like Random Dopant Fluctuations (RDF) and systematic spatially-correlated effects like dose, focus and overlay variations [38]. These uncertainties impact various device characteristics, such as effective gate length, oxide thickness and transistor threshold voltages. Altered device characteristics may lead to significant variations in power consumption and to timing violations. While PV has been addressed extensively in the circuit and architecture communities, its impact in NoC architectures remains largely unexplored despite the menacing signs.

Given the looming prominence of both reliability and variability artifacts in deep sub-micron circuits, it is no longer sufficient to concentrate solely on the triptych of ***Performance/Area/Power***. Fourth and fifth pillars – comprising ***Reliability/Variability*** issues – should be added to the above trio of evaluation metrics to create a quintet of fundamental design drivers. This penta-faceted approach to evaluating NoC architectures forms the premise of the research presented in this volume.

The work described hereafter aims at developing *a comprehensive framework for designing high-performance, area- and energy-efficient, reliable, thermal-aware, and PV-resilient on-chip networks*, considering a multi-dimensional design space and the underlying technology constraints. Toward that extent, the said research takes a *holistic* approach encompassing the interplay of the previously mentioned evaluation metrics and their synergistic and symbiotic effects. The primary intellectual focus of this endeavor is to *understand the tradeoffs and corresponding ramifications of designing mission-critical components destined for severely constrained environments*, while still satisfying all the prescribed requirements.

1.5 Overview of Research

The previous section outlined the significance of reliability and variability in the design and evaluation of on-chip networks. The research presented in this volume will, therefore, give equal weighting to all five design metrics, and attempt to provide a detailed design space exploration of NoC architectures encompassing all primary design drivers. The aim is to develop architectural solutions to tackle all issues involved in a modern interconnection fabric connecting multiple cores. The problem will be attacked through a *two-pronged process*: (1) ***MICRO-Aarchitectural Innovations*** at the granularity of individual hardware components, and (2) ***MACRO-Architectural Solutions*** at a higher abstraction level; i.e., looking at the interconnection system as a whole. These two threads are highly interdependent and cannot be viewed in isolation; they complement each other and are entangled in an elaborate interplay which encircles the entire design. Efficient NoC implementations should address both facets and ensure seamless integration of the two. The proposed approach is illustrated abstractly in Fig. 1.3. As indicated in the figure, the micro-architectural investigation will concentrate on (a) the on-chip ***Router*** and (b) the ***Network Interface Controller (NIC)***, which constitute the main components within a network node. The bulk of the exploration will focus on the router, which offers a lot more flexibility in design choices. The macro-architectural

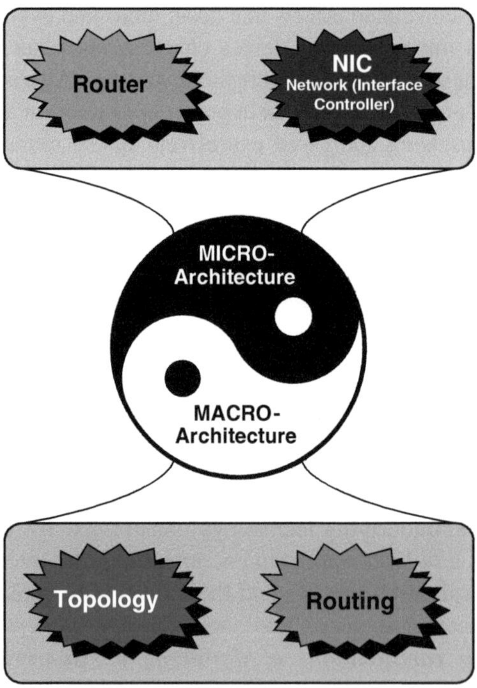

Fig. 1.3 High-level overview of the proposed research exploration

investigation will address (a) the network ***Topology*** and (b) the ***Routing*** process and associated algorithms.

Figures 1.4 and 1.5 expand on Fig. 1.3, and provide a more detailed overview for each of the two proposed paths (the micro- and macro-architectural approaches). The figures summarize the work completed by the author in the major design categories of the two core themes. Individual research projects are identified by a design keyword and the conference at which they appeared (see legend on page 9). The grid at the bottom of both figures classifies the projects based on their contribution to the previously mentioned quintet of fundamental evaluation metrics: (1) Performance, (2) Power, (3) Area, (4) Reliability, and (5) Variability.

Figure 1.4 (MICRO) divides the Router theme into its main components (***Buffers, Arbiters,*** and ***Crossbar***), while the NIC is investigated in terms of its main characteristic, i.e., the architectural ***Structure***. Similarly, Fig. 1.5 (MACRO) divides the Topology theme into ***Hybridization*** and ***Links***, and the Routing theme into ***Deterministic*** and ***Adaptive***. All these sub-categories combine to provide a complete picture of the NoC architecture, ranging from fine-grained (MICRO) to coarser-grained (MACRO) granularities.

1.5.1 *Legend for Figs. 1.4 and 1.5*

ViChaR: **"ViChaR: A Dynamic Virtual Channel Regulator for Network-on-Chip Routers,"** presented at the 39th Annual International Symposium on Microarchitecture (MICRO), December 2006 [39].

RoCo: **"A Gracefully Degrading and Energy-Efficient Modular Router Architecture for On-Chip Networks,"** presented at the 33rd Annual International Symposium on Computer Architecture (ISCA), June 2006 [40].

[AC,

PA,

TPA,

SON,

HBH Scheme]: **"Exploring Fault-Tolerant Network-on-Chip Architectures,"** presented at the International Conference on Dependable Systems and Networks (DSN), June 2006 [37].

ANCS-05: **"Design and analysis of an NoC architecture from performance, reliability and energy perspective,"** presented at the Symposium on Architecture for Networking and Communications Systems (ANCS), October 2005 [41].

MEP: **"A Distributed Multi-Point Network Interface for Low-Latency, Deadlock-Free On-Chip Interconnects,"** presented at the International Conference on Nano-Networks (Nano-Net), September 2006 [42].

3D NetInMem: **"Design and Management of 3D Chip Multiprocessors Using Network-in-Memory,"** presented at the 33rd Annual International Symposium on Computer Architecture (ISCA), June 2006 [43].

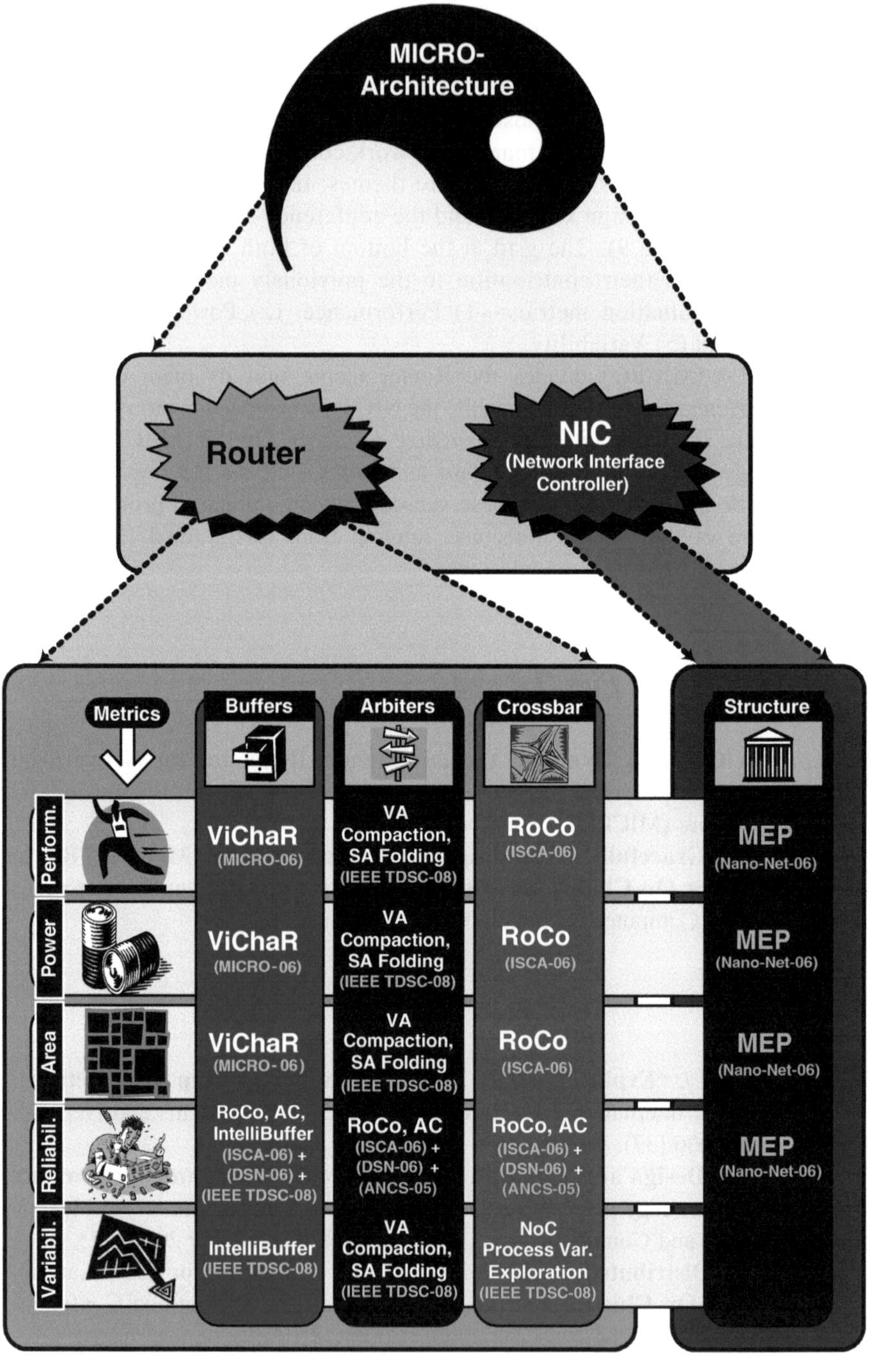

Fig. 1.4 Overview of MICRO-architectural research (see Section 1.5.1 for legend)

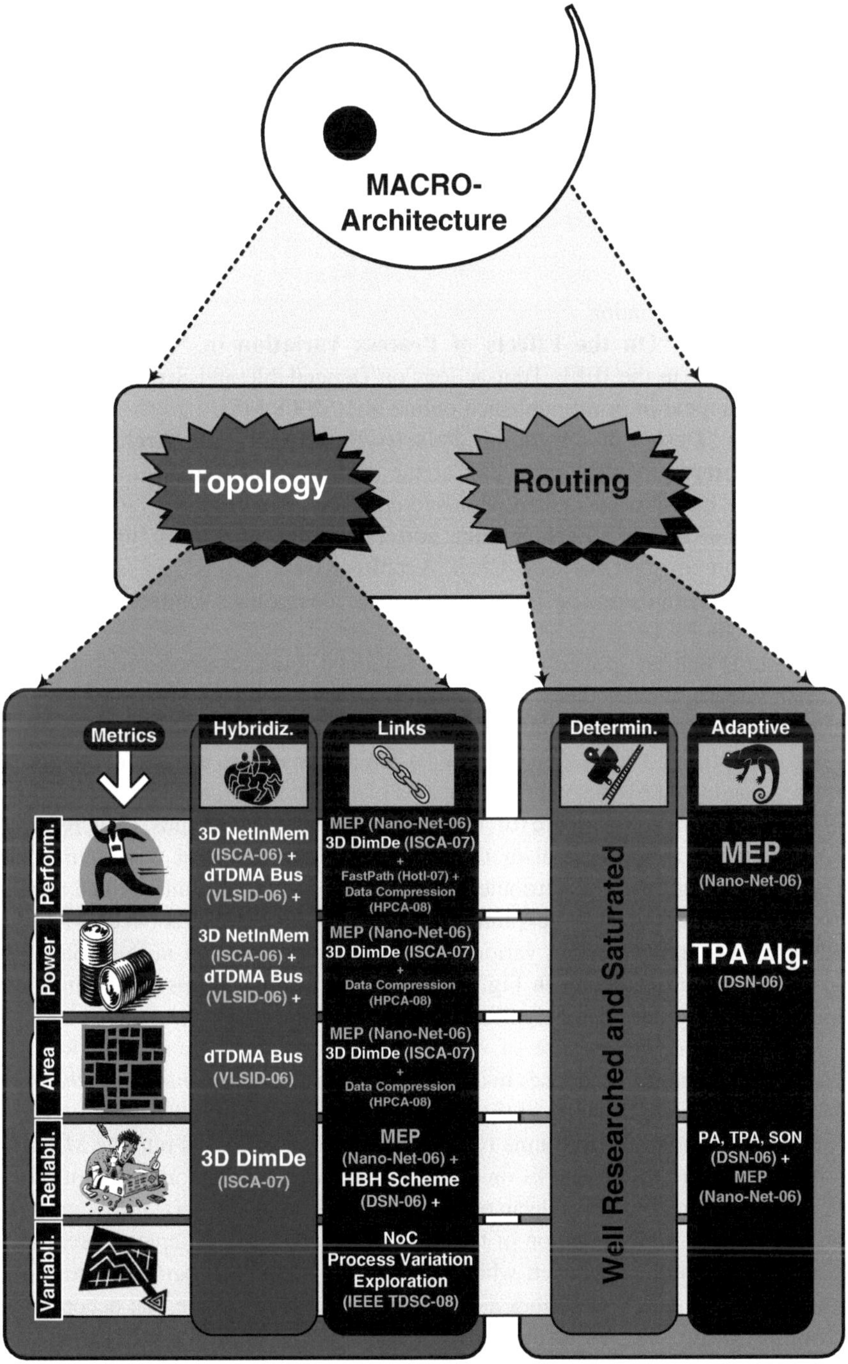

Fig. 1.5 Overview of MACRO-architectural research (see Section 1.5.1 for legend)

3D DimDe: **"A Novel Dimensionally-Decomposed Router for On-Chip Communication in 3D Architectures,"** presented at the 34th Annual International Symposium on Computer Architecture (ISCA), June 2007 [44].

dTDMA Bus: "**A Hybrid SoC Interconnect with Dynamic TDMA-Based Transaction-Less Buses and On-Chip Networks,**" presented at the 19th International Conference on VLSI Design, January 2006 [15].

[IntelliBuffer,
VA Compaction,
SA Folding,
NoC Process Variation
Exploration]: **"On the Effects of Process Variation in Network-on-Chip Architectures,"** in the IEEE Transactions on Dependable and Secure Computing (TDSC), to appear in print; published online in 10.2008 [45].

FastPath: **"Design of a Dynamic Priority-Based Fast Path Architecture for On-Chip Interconnects,"** presented at the 15th Annual Symposium on High-Performance Interconnects (IEEE Hot Interconnects), August 2007 [46].

Data Compression: **"Performance and Power Optimization through Data Compression in Network-on-Chip Architectures,"** presented at the 14th International Symposium on High-Performance Computer Architecture (HPCA), February 2008 [47].

Keywords that are grayed-out in Figs. 1.4 and 1.5 indicate projects that do not form an integral part of this volume; however, they are included because of their peripheral significance and complementary role to the underlying concepts. These projects are briefly discussed in Chapter 10, which constitutes a digest of additional research conducted by the author in the area of on-chip interconnects.

The complementary nature of the two core thematic strands (MICRO and MACRO) implies that several of the projects contained within this volume fall under both categories. This attribute is further testament to the intertwined ramifications of NoC architectural exploration. The apparent overlap between MICRO- and MACRO- architecture in various pieces of work presented in this volume is illustrated diagrammatically in Fig. 1.6. In this figure, keywords shown in white represent work of peripheral significance to this volume.

Building upon the premise of exploring two research streams (MICRO and MACRO) this volume comprises *two distinct sections, each spanning four chapters*. The overall thematic structure of the volume is illustrated abstractly in Fig. 1.7.

The *first section* of the volume (Chapters 3 through 6) delves into the MICRO-architectural world of Networks-on-Chip; it contains detailed work on most of the major components found inside an on-chip router, as well as several novel concepts enhancing the overall operation of the network. Specifically, Chapter 3 presents a novel dynamic buffer manager, which optimizes the size and performance of the NoC buffers. Chapter 4 describes a decoupled router architecture that decomposes incoming traffic into two independent flows based on the direction of travel. The two flows are subsequently served by two smaller, leaner, and faster modules that are independently operated; this allows for partial router operation and graceful degradation. Chapter 5 develops several new ideas on fault-tolerance and reliability

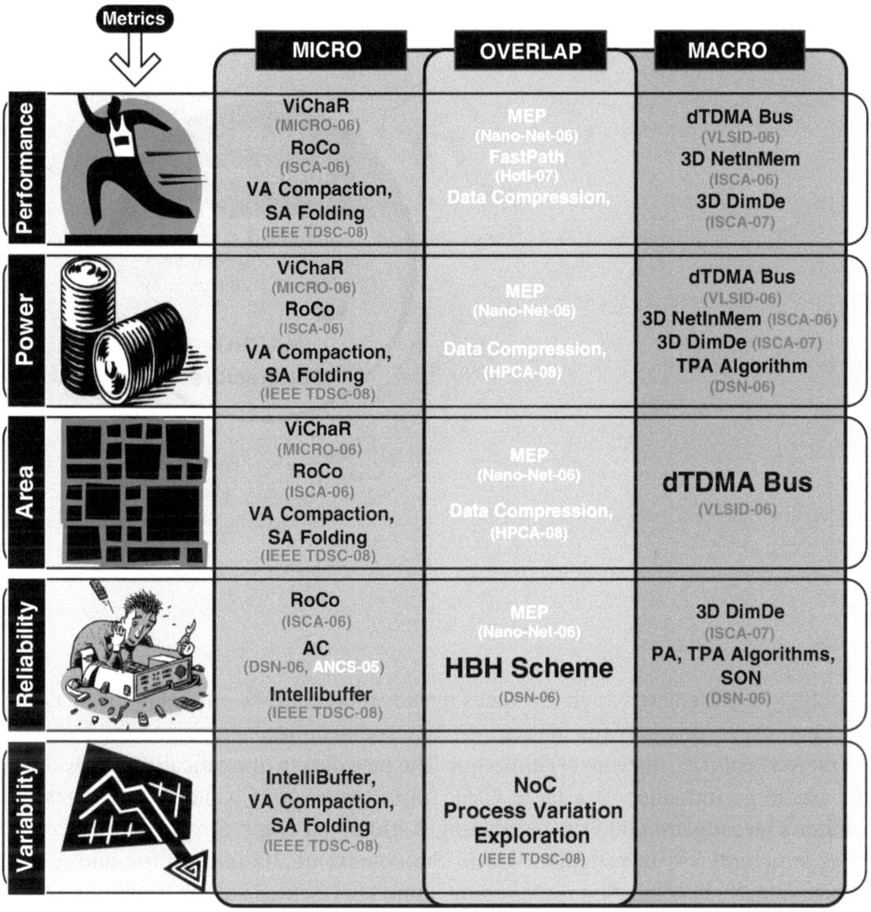

Fig. 1.6 Overlap between MICRO- and MACRO-architectural NoC research

in the presence of both soft (i.e., transient) and hard (i.e., permanent) failures in either the inter-router links or within the routers themselves. Finally, Chapter 6 tackles the relatively new and rapidly emerging threat of Process Variation. Several architectural solutions are proposed to account for process variability and counter its adverse effects.

The *second section* of this volume (Chapters 7 through 10) shifts focus to the MACRO-architecture of the NoC. Attention now is turned to higher abstraction layers that view the network as a whole system and try to investigate the behavior of both the network and the processing elements as a single entity working in unison. Chapter 7 starts this exploration by investigating the notion of interconnect hybridization; a newly developed bus structure is fused with an underlying NoC to reap benefits from both architectures in a unified environment. Chapter 8 extends the notion of hybridization to the third dimension. A new three-dimensional network

Fig. 1.7 Thematic structure of volume

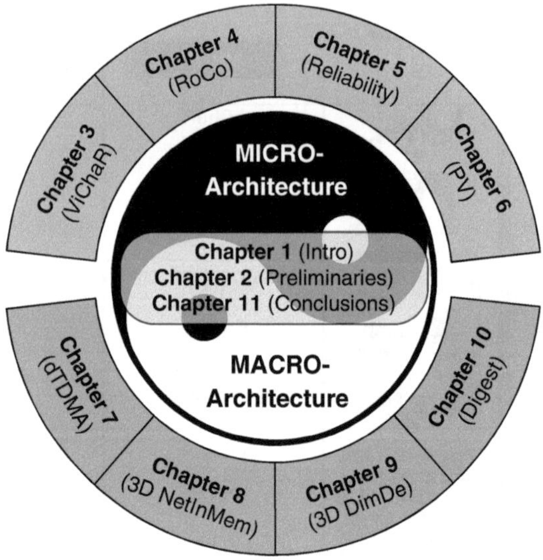

topology is presented, which leverages the advantages of the emerging 3D stacking technology. In this implementation, the bus architecture is employed for transfers in the vertical (i.e., inter-layer) direction. The new design dramatically improves the L2 cache performance of a large CMP implementation. Chapter 9 continues the author's investigation into the fascinating world of 3D chips. Several new interconnect structures are investigated within the context of 3D integration, and a new dimensionally decomposed router architecture is presented, which leverages a true, physical 3D crossbar design. This new architecture constitutes a natural extension of the 2D decoupled router of Chapter 4. Finally, Chapter 10 presents a digest of additional research conducted by the author in the field of NoCs. This work is presented in summary, because it was not deemed to be of integral value to this volume. However, the salient features of these projects are certainly complementary and of significant peripheral value to the overall volume.

The work presented in this volume address the issues under investigation through the aforementioned penta-faceted evaluation prism, which encompasses the five key metrics of (1) **performance**, (2) **silicon area consumption**, (3) **power/energy efficiency**, (4) **reliability**, and (5) **variability**.

Chapter 2
A Baseline NoC Architecture

A generic NoC implementation consists of a number of Processing Elements (PE) arranged in a mesh-like grid, as shown in Fig. 2.1 The PEs may be of the same type, e.g., CPU, or of different type, e.g., audio cores, video cores, wireless transceivers, memory banks, etc. Each PE is connected to a local router through a Network Interface Controller (NIC); each router is, in turn, connected to adjacent routers forming a packet-based on-chip network. The NIC module packetizes/de-packetizes the data into/from the underlying interconnection network. The PE together with its NIC form a network node. Nodes communicate with each other by injecting data packets into the network. The packets traverse the network toward their destination, based on various routing algorithms and control flow mechanisms.

The heart of an on-chip network is the router, which undertakes the crucial task of steering and coordinating the data flow. The architecture employed by conventional NoC routers [48] is illustrated in Fig. 2.2 The router operation revolves around two fundamental regimes: (a) the datapath, and (b) the associated control logic.

The datapath consists of a number of input and output channels to facilitate packet switching and traversal. In general, the router has P input and P output channels (or ports). In most implementations, P = 5; four inputs from the four cardinal directions (North, East, South and West) and one from the local Processing Element (PE), which is attached to the NoC router. To minimize router complexity and traffic congestion, NoC routers are usually assumed to connect to a single PE. The input/output channels may consist of unidirectional links (as shown in Fig. 2.2), bidirectional, or even serial links.

Buffering within a network router is necessary due to congestion, output link contention, and intra-router processing delays (e.g., routing computation), which impede data flow. In the case of virtual channel-based NoC routers, each input port consists of a number of FIFO buffers, with each FIFO corresponding to a virtual channel (see Fig. 2.2). Hence, each input port has v virtual channels, each of which has a dedicated k-flit FIFO buffer (a flit is the smallest unit of flow control; one network packet is composed of a number of flits). Given the very limited buffer space in resource-constrained on-chip networks, routers tend to employ wormhole flow control, which relaxes the constraints on buffer size as compared to store-and-forward and virtual cut-through.

C. Nicopoulos et al., *Network-on-Chip Architectures: A Holistic Design Exploration*, Lecture Notes in Electrical Engineering 45,
DOI 10.1007/978-90-481-3031-3_2, © Springer Science+Business Media B.V. 2009

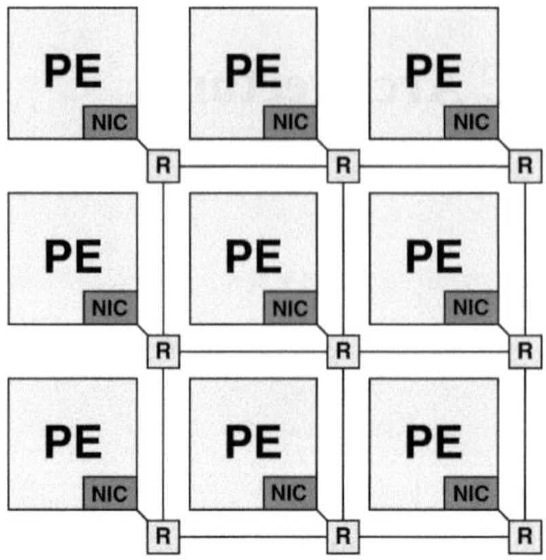

Fig. 2.1 A generic NoC architecture (3×3 mesh)

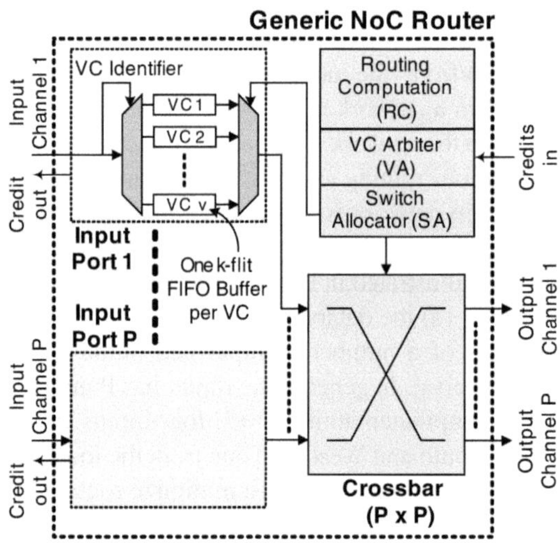

Fig. 2.2 A conventional NoC router

The router control logic forms the heart of the NoC router, and comprises four components: (a) the Routing Computation (RC) unit, (b) the Virtual Channel Arbitration (VA) logic, (c) the Switch Allocation (SA) logic, and (d) the Crossbar (XBAR).

The RC unit is responsible for directing the header flit of an incoming packet to the appropriate output Physical Channel (PC) and/or dictating valid output Virtual Channels (VC) within the selected PC. Output VCs are, essentially, the input VCs of the adjacent routers. The routing is done based on the packet's destination address, which is present in the header flit, and can be deterministic (e.g., dimension-order routing) or adaptive (e.g., load-balancing). RC is a "per-packet" operation: it is performed once for each packet within a router (i.e., only on the header flit of each packet).

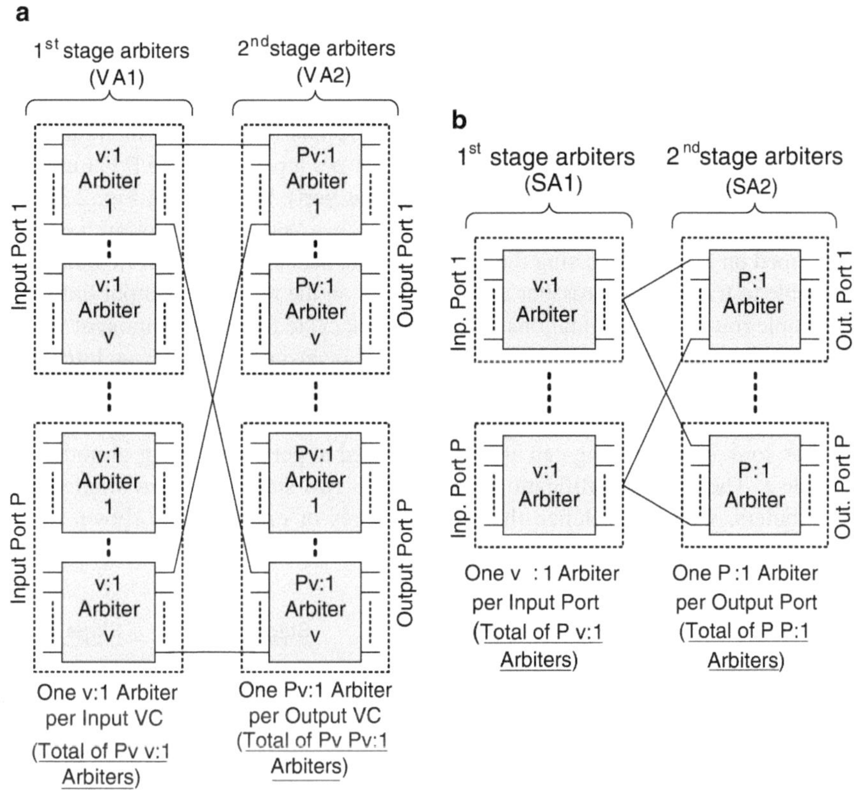

Fig. 2.3 NoC router arbitration/allocation logic. (**a**) VC arbitration (VA). (**b**) Switch allocation (SA)

The Virtual Channel Arbitration (VA) module arbitrates amongst all packets requesting access to the same VCs and decides on winners. Since the routing function may not specify a particular output VC in the requested output physical port [49], two arbitration stages are generally required, as shown in Fig. 2.3a. The first stage (VA1) reduces the number of requests from each input VC to one; this ensures the request of a single VC at a particular output port by each input VC. A total of Pv v:1 arbiters are, therefore, required in the first arbitration stage (see Fig. 2.3a): one arbiter for each input VC. Subsequently, the winning request from each input VC proceeds to the second arbitration stage (VA2). A total of Pv Pv:1 arbiters are required in this stage: one arbiter for each output VC (each arbiter is of Pv:1 size to accommodate the worst-case scenario in which all input VCs request the same output VC). Just like RC, VA is also a "per-packet" operation; it is only performed on header flits.

The SA unit arbitrates amongst all VCs requesting access to the crossbar and grants permission to the winning flits. Switch allocation is also performed in two stages. The first – local – stage (SA1) accounts for the sharing of a single port by a number of VCs; all VCs within the same input port compete against each other locally for access to the output physical channels. The second – global – stage (SA2) arbitrates between the winning requests from each input port for each output port. The SA2 stage sets the crossbar control signals accordingly. The SA unit is very similar in structure to the VA unit. It consists of logically identical arbiters arranged in cascaded fashion. The difference lies only in the size and number of arbiters used: P v:1 arbiters are required for SA1 (i.e., one arbiter per input port), and P P:1 arbiters are required for SA2 (i.e., one arbiter per output port), as shown in Fig. 2.3b. As opposed to the VA's "per-packet" operation, SA is a "per-flit" operation, i.e., it is performed on all flits traversing the router, not just header flits. The SA winners are then able to traverse the crossbar and are placed on the respective output links.

Simple router implementations require a clock cycle for each component within the router, as illustrated at the top of Fig. 2.4. Lower-latency router architectures parallelize the VA and SA using speculative allocation, which predicts the winner of the VA stage and performs SA based on that [49] (3-stage Router in Fig. 2.4). Further, look-ahead routing can also be employed to perform routing of node $i+1$ at node i. These two modifications have led to two-stage and even single-stage [48] routers, which parallelize the various stages of operation, as shown in the lower half of Fig. 2.4.

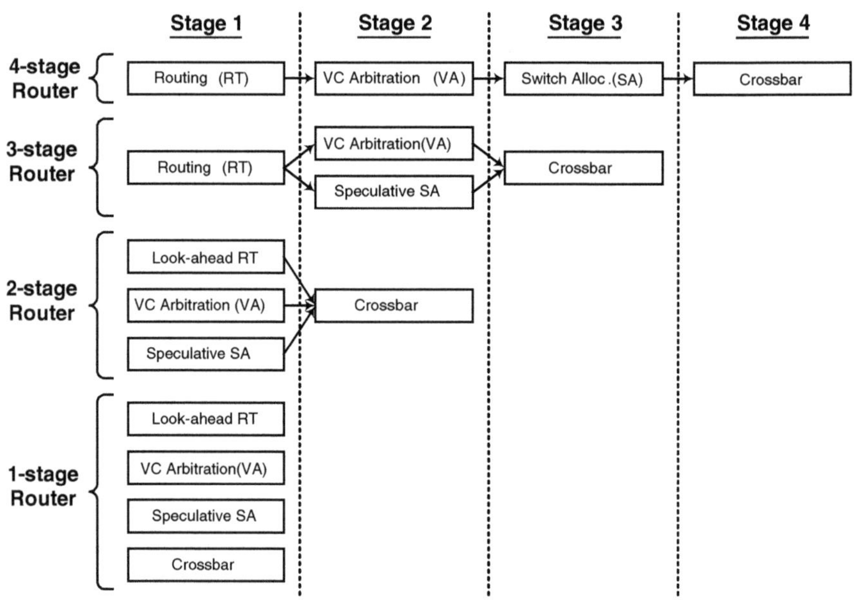

Fig. 2.4 Various NoC router pipeline implementations

Part I
MICRO-Architectural Exploration

Chapter 3
ViChaR: A Dynamic Virtual Channel Regulator for NoC Routers [39]

Router buffers are instrumental in the overall operation of the on-chip network. Besides performance, buffers greatly affect the network's overall energy budget. In fact, of the different components comprising the interconnection fabric of SoCs, buffers are the largest leakage power consumers in an NoC router, consuming about 64% of the total router leakage power [50]. Similarly, buffers consume significant dynamic power [22,51] and this consumption increases rapidly as packet flow throughput increases [51]. In fact, it has been observed that storing a packet in a buffer consumes far more energy than transmitting the packet [51]. Furthermore, the area occupied by an on-chip router is dominated by the buffers [14,52,53]. Consequently, buffer design plays a crucial role in architecting high performance and energy efficient on-chip interconnects, and is the focus of this section.

This chapter introduces a novel unified buffer structure – the dynamic Virtual Channel Regulator (ViChaR) – which dynamically allocates buffer resources according to network conditions.

3.1 Importance of Buffer Size and Organization

Decreasing the buffer size arbitrarily to reclaim silicon area and minimize power consumption is not a viable solution, because of the intricate relationship between network performance and buffer resources. Buffer size and management are directly linked to the flow control policy employed by the network; flow control, in turn, affects network performance and resource utilization. Whereas an efficient flow control policy enables a network to reach 80% of its theoretical capacity, a poorly implemented policy would result in a meager 30% [54]. Wormhole flow control [55] was introduced to improve performance through finer-granularity buffer and channel control at the flit level instead of the packet level (a flit is the smallest unit of flow control; one packet is composed of a number of flits). This technique relaxes the constraints on buffer size at each router, allowing for a more efficient use of storage space than store-and-forward and virtual cut-through [56] switching. However, the channel capacity is still poorly utilized; while the buffers

C. Nicopoulos et al., *Network-on-Chip Architectures: A Holistic Design Exploration*, Lecture Notes in Electrical Engineering 45,
DOI 10.1007/978-90-481-3031-3_3, © Springer Science+Business Media B.V. 2009

are allocated at the flit level, physical paths are still allocated at the packet level. Hence, a blocked packet can impede the progress of other packets waiting in line and may also cause multi-node link blocking (a direct consequence of the fact that the flits of a single packet are distributed across several nodes in wormhole routers). To remedy this predicament, Virtual Channel (VC) flow control [57] assigns multiple virtual paths (each with its own associated buffer queue) to the same physical channel. It has been shown that VC routers can increase throughput by up to 40% over wormhole routers without VCs [54]. As a side bonus, virtual channels can also help with deadlock avoidance [58]. The work in this report assumes, without loss of generality, the use of VC-based wormhole flow control, which suits the low buffer requirements of NoC routers.

The way virtual channels – and hence buffers – are organized within a router is also instrumental in optimizing performance. The number of VCs per physical channel and the VC depth are two parameters that form an elaborate interplay between buffer utilization, throughput and latency. Researchers in the macro-network field have identified the decisive role of virtual channel organization in overall system performance [59,60]. Detailed studies of the relation between virtual channels and network latency indicate that for low traffic intensity, a small number of VCs can suffice. In high traffic rates, however, increasing the number of VCs is a more effective way of improving performance than simply increasing the buffer depth [60]. Under light network traffic, the number of packets traveling through a router is small enough to be accommodated by a limited number of VCs; increasing the number of VCs yields no tangible benefits. Under high traffic, many packets are contenting for router resources; increasing VC depth will not alleviate this contention because of Head-of-Line (HoL) blocking. Increasing the number of VCs, though, will allow more packets to share the physical channels. This dichotomy in VC organization implies that routers with fixed buffer structures will either be underutilized or will underperform under certain traffic conditions, as illustrated in the examples of Fig. 3.1 This figure highlights the weaknesses of statically-partitioned buffers.

Since buffer resources come at a premium in resource-constrained NoC environments (they consume valuable power and silicon real-estate), it is imperative to limit the buffer size to a minimum without severely affecting performance. This objective function can only be achieved through the use of efficient management techniques which optimize buffer utilization. Since size and organization are design-time decisions, they cannot be dynamically changed during operation based on observed traffic patterns. However, the use of a carefully designed buffer controller can significantly affect the efficiency of storing and forwarding of the flits. Therefore, the throughput of a switch can be maximized through dynamic and real-time throttling of buffer resources.

Given the aforementioned significance of the NoC buffers in the area, power and performance triptych, we thereby introduce ViChaR[1]: a dynamic Virtual Channel Regulator, which dispenses VCs according to network traffic. The ViChaR module

[1] The name ViChaR was intentionally chosen to echo the word Vicar, who is someone acting as a substitute or agent for a superior.

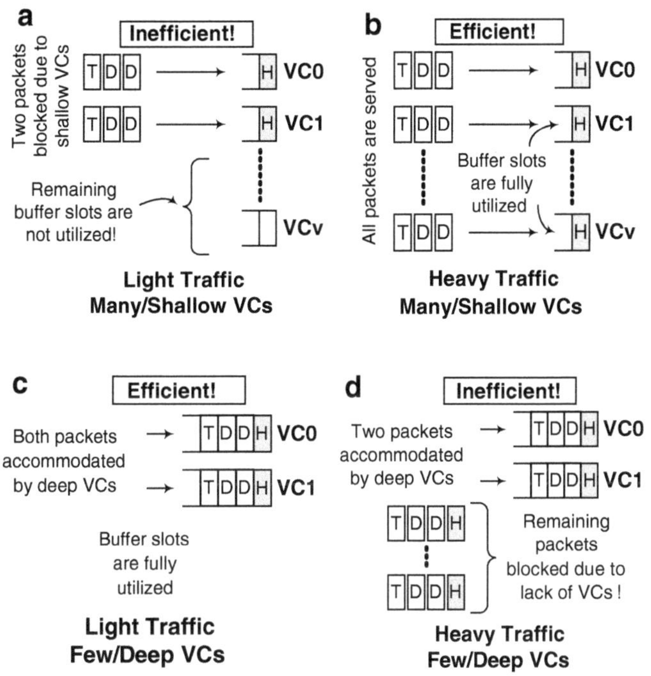

Fig. 3.1 Limitations of a statically assigned buffer organization (H = head flit, D = data/middle flit, T = tail flit). (**a**) Light traffic many/shallow VCs. (**b**) Heavy traffic many/shallow VCs. (**c**) Light traffic few/deep VCs. (**d**) Heavy traffic few/deep VCs

is a very compact unit operating at the granularity of one router input/output port; therefore, a conventional 5-port NoC router would employ five such units to oversee buffer management.

ViChaR's operation revolves around two intertwined concepts which constitute the *two fundamental contributions* of this work:

(1) ViChaR Uses a Unified Buffer Structure (UBS), instead of individual and statically partitioned First-In-First-Out (FIFO) buffers. While the unified buffer concept is not new, in this work we are revisiting the concept within the confinements of the strict resource limitations of on-chip networks. This is the first attempt to incorporate a unified buffer management in NoC routers. The new flavor in our endeavor stems from a fundamentally different implementation approach: we introduce a novel, table-based design that provides single-clock operation without incurring prohibitive overhead. Most importantly though, it enables the use of a flexible and dynamically varying virtual channel management scheme, thereby replacing the conventional static resource allocation.
(2) ViChaR provides each individual router port with a variable number of VCs, each of which is dispensed dynamically according to network traffic conditions. This translates into fewer but deeper VCs under light traffic, and more but shallower VCs under heavy traffic. This attribute successfully marries two

contradicting buffer organizations, which are impossible to combine in conventional, statically-allocated buffers. Furthermore, ViChaR's dynamic allocation scheme ensures a smooth continuum between these two extremes (few/deeper VCs versus more/shallower VCs) as the network intensity fluctuates.

3.2 Related Work in Buffer Design

As previously mentioned, interest in packet-based on-chip networks has rapidly gained momentum over the last few years, and analysis and optimization of on-chip interconnect architectures have garnered great attention. In this sub-section, we focus solely on buffer related aspects. Within the realm of on-chip buffer design, both size and organization have been shown to be directly related to network performance [52]. Buffer sizing in particular has been investigated in [52,53]. However, these papers adopt a static approach, where optimal buffer sizes are predetermined at design-time based on a detailed analysis of application-specific traffic patterns. The sizing is optimal for only one particular application and one hardware mapping. However, a technique to alter the buffer organization dynamically at run-time is more desirable for a general purpose and reconfigurable SoC executing different workloads. A dynamic scheme would maximize utilization regardless of the traffic type in the NoC.

Centralized buffer organizations have been studied extensively in the macro-network realm, but the solutions proposed are not amenable to resource constrained on-chip implementations. In particular, a unified and dynamically-allocated buffer structure was originally presented in [61] in the form of the Dynamically Allocated Multi-Queue (DAMQ) buffer. However, whereas the DAMQ architecture was part of a single-chip communication coprocessor for multi-computer systems, the proposed implementation in this report is aimed at area- and power-constrained, ultra-low latency on-chip communication. This profoundly affected our design considerations as follows:

(1) The DAMQ used a fixed number of queues (i.e., virtual channels) per input port. Specifically, four queues were used, one for each of three output ports and a local processor interface. Consequently, all packets in the same queue had to obey the FIFO order, i.e., all packets in the same queue could still get stuck behind a blocked packet at the head of the queue.
(2) The control logic of the DAMQ buffer was very complex, relying on a system of linked lists to organize the data path. These linked lists were stored in pointer registers which had to be updated constantly. This caused a three-cycle delay for every flit arrival/departure, mainly because data had to be moved between pointer registers, and a so-called "free list" had to be updated (a linked list keeping track of available buffer slots) [62]. This three-cycle delay – while acceptable for inter-chip communication – would prove intolerable in an on-chip router.

The DAMQ project spawned a few other designs, which aimed to simplify the hardware implementation and lower overall complexity. Two notable examples of these designs were the DAMQ with self-compacting buffers [63] and the Fully Connected Circular Buffer (FC-CB) [64]. Both designs have less overhead than the linked-list approach of [61] by employing registers, which selectively shift some flits inside the buffer to enable all flits of one VC to occupy a contiguous buffer space. The FC-CB design [64] improves on [63] by using a circular structure, which shifts in only one direction and ensures that any flit will shift by at most one position each cycle. However, the FC-CB has two main disadvantages when applied to an on-chip network. First, being fully connected, it requires a $P^2 \times P$ crossbar instead of the regular $P \times P$ crossbar for a P-input switch. Such large and power-hungry crossbars are unattractive for on-chip routers. Second, the circular shifter allows an incoming flit to be placed anywhere in the buffer and requires selective shifting of some parts of the buffer while leaving the rest of the buffer undisturbed. This functionality inflicts considerable increases in latency, area and power over a simple, non-shifting buffer implementation, like the proposed ViChaR design. The overhead is due to the large MUXes, which are required between each buffer slot to enable both shifting and direct input.

The circular-shift buffer of the FC-CB was implemented in Verilog HDL and synthesized in 90 nm commercial TSMC libraries to assess its viability in on-chip designs. The circular buffer implementation of the FC-CB increases the datapath delay by 26% compared to ViChaR's stationary (i.e., non-shifting) buffer. Increases in datapath delay may affect the pipeline period in deeply pipelined router designs, a longer period will adversely affect throughput. Moreover, the FC-CB's large MUXes incur an increase of approximately 18% in buffer area. More importantly, though, the continuous shifting of the FC-CB buffer every clock cycle (assuming continuous incoming traffic) increases the dynamic power budget by 66%. Obviously, this overhead renders the FC-CB implementation unattractive for on-chip applications. Finally, the FC-CB still works with a fixed number of VCs, just like the DAMQ design. In this report, we will show that a dynamically variable number of VCs optimizes performance.

The notion of dynamically allocating VC resources based on traffic conditions was presented in [65], through the VCDAMQ and DAMQ-with-recruit-registers (DAMQWR) implementations. However, both designs were coupled to DAMQ underpinnings; hence, they employed the linked-list approach of the original DAMQ, which is too costly for an on-chip network. Nevertheless, the work of [65] highlighted the significance of dynamic allocation of buffer resources, which forms the premise of the design proposed in this report.

Finally, the Chaos router [66] and BLAM routing algorithm [67] provide an alternative technique to saving buffer space. They employ packet misrouting, instead of storage, under heavy load. However, randomized (non-minimal) routing may make it harder to meet strict latency guarantees required in many NoCs (e.g., multimedia SoCs). Moreover, these schemes do not support dynamic VC allocation to handle fluctuating traffic.

3.3 The Proposed Dynamic Virtual Channel Regulator (ViChaR)

So far, as a result of scarce area and power resources and ultra-low latency requirements, on-chip routers have relied on very simple buffer structures. In the case of virtual channel-based NoC routers, these structures consist of a specified number of FIFO buffers per input port, with each FIFO corresponding to a virtual channel. This is illustrated in Fig. 3.2 Such organization amounts to a static partitioning of buffer resources. Hence, each input port of an NoC router has *v* virtual channels, each of which has a dedicated *k*-flit FIFO buffer. Current on-chip routers have small buffers to minimize their overhead; *v* and *k* are usually much smaller than in macro networks [48]. The necessity for very low latency dictates the use of a parallel FIFO implementation, as shown in the bottom right of Fig. 3.2 As opposed to a serial FIFO implementation [68], the parallel flavor eliminates the need for a flit to traverse all slots in a pipelined manner before exiting the buffer [68]. This fine-grained control requires more complex logic, which relies on read and write pointers to maintain the FIFO order. Given the small sizes of on-chip buffers, though, the inclusion of a parallel FIFO implementation is by no means prohibitive. The buffers within an NoC router can be implemented as either registers or SRAM/DRAM memory [69,70]. However, given the relatively small buffer sizes employed, it is more reasonable to use small registers as buffers to avoid the address decoding/ encoding latencies of big memories and the access latencies associated with global bitlines/wordlines [69]. To this extent, the NoC buffers in this report were implemented as registers.

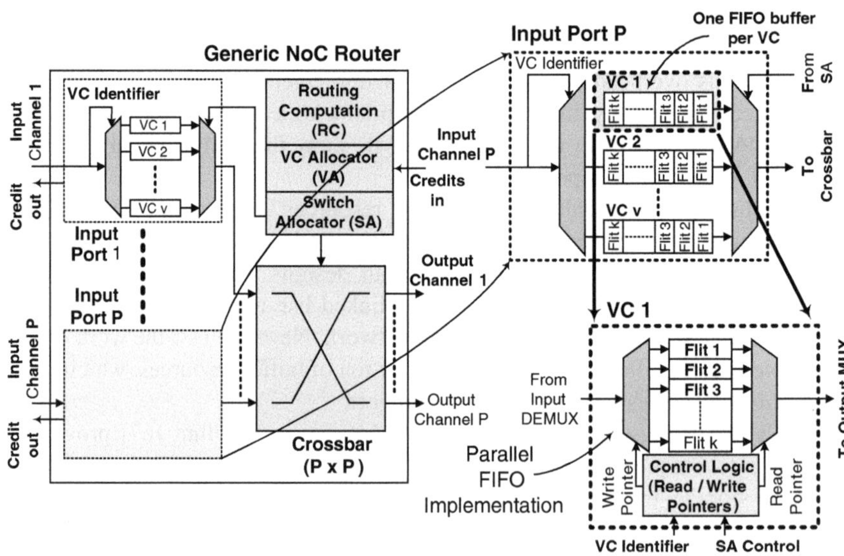

Fig. 3.2 A generic NoC router architecture and buffer structure

FIFO buffers in statically assigned buffer structures have two inherent disadvantages. First, a packet at the head of a VC whose designated output port is busy will block all subsequent packets in that VC from being transmitted (assuming non-atomic buffer allocation) even if their designated output ports are free. This Head-of-Line (HoL) blocking can severely affect network performance in congested conditions, similar to the previously discussed DAMQ. This scenario is illustrated at the top of Fig. 3.3 Second, if only part of a packet occupies a VC buffer at a given time, then any vacant slots in that buffer cannot be reassigned to a new packet for as long as that VC is reserved by the partial packet to avoid packet/message mixing. Thus, a VC buffer may only be occupied by a single header flit because the remaining flits happen to be blocked in preceding routers due to congestion. In such a scenario, the remaining free slots in the buffer cannot be assigned to other packets until the tail flit of the current packet releases the VC. This attribute of FIFO buffers can lead to substantial under-utilization of the buffers, as shown at the bottom of Fig. 3.3 and cripple network performance.

Figure 3.4 illustrates the buffer organization of a conventional NoC router (left) and our proposed alterations (right). The crux of ViChaR is composed of two main components: (1) the ***Unified Buffer Structure (UBS)***, shown in Fig. 3.4, and (2) the associated control logic, called ***Unified Control Logic (UCL)***.

Figure 3.4 shows only one of the five sub-modules of UCL, the *Arriving/Departing Flit Pointer Logic*. This sub-module constitutes the interface between the UBS and the UCL; the UCL controls the unified buffer (UBS) through the Arriving/Departing Flit Pointer Logic module. A top-level block diagram of the entire ViChaR architecture is shown in Fig. 3.6. This figure illustrates all five of the UCL sub-modules: (1) the *Arriving/Departing Flit Pointers Logic*, (2) the *Slot Availability Tracker*, (3) the *VC Availability Tracker*, (4) the *VC Control Table*, and (5) the *Token (VC) Dispenser*. The operation of each component and the interaction

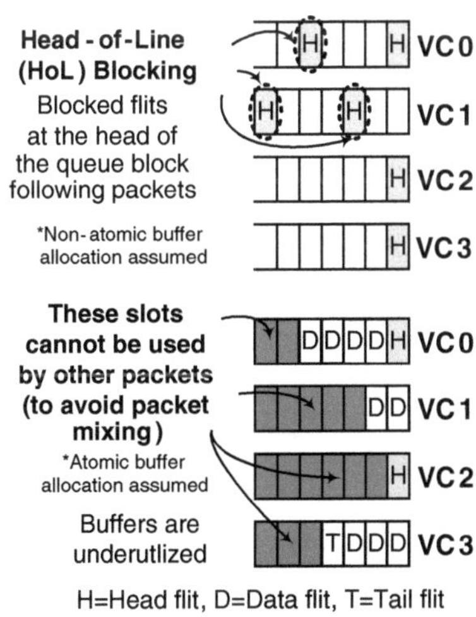

Fig. 3.3 Limitations of existing FIFO buffers

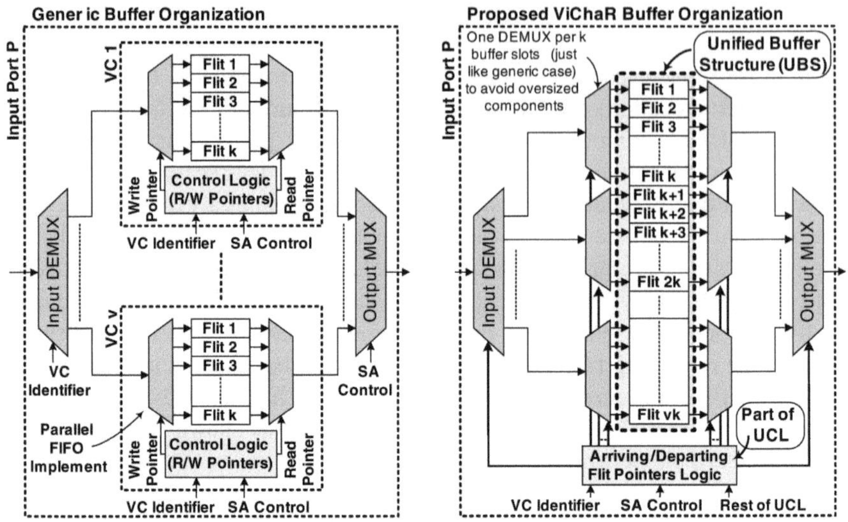

Fig. 3.4 The proposed ViChaR architecture

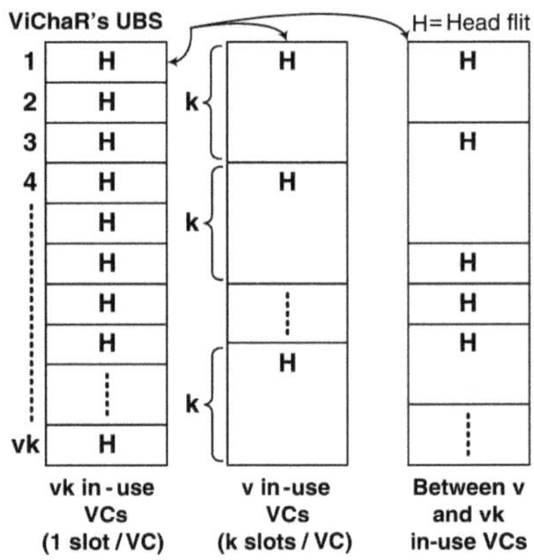

Fig. 3.5 Possible VC configurations in ViChaR

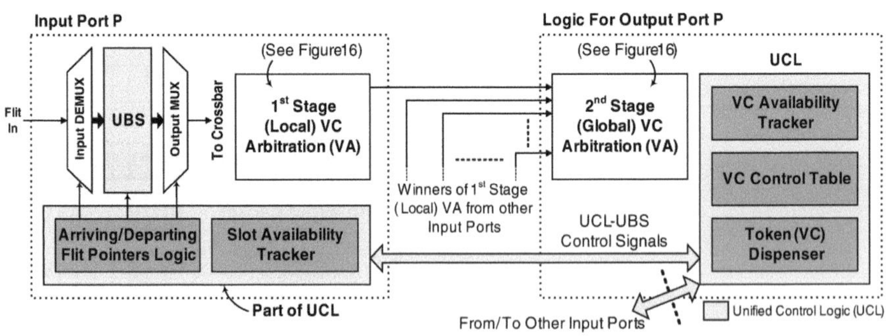

Fig. 3.6 ViChaR block diagram (only one of P ports shown here)

between the UBS and its controlling entity (the UCL) are described in detail in Section 3.3.2. All five modules function independently and in parallel, which is of critical importance to the ultra-low latency requirements of the router. The UCL components work in tandem with the unified buffer (UBS), providing dynamic allocation of both virtual channels and their associated buffer depth. As illustrated in Fig. 3.6, the two main ViChaR components (UBS and UCL) are logically separated into two groups: the unified buffer (UBS) and two of the five UCL modules (the Arriving/Departing Flit Pointers Logic and the Slot Availability Tracker) are situated at the input side of the router (i.e., to accept all incoming flits), while the remaining modules of the control logic (UCL) are responsible for the VC arbitration of all flits destined to a particular output port. Based on incoming traffic and information from the Slot and VC Availability Trackers, thc Token (VC) Dispenser grants VC IDs to new packets accordingly. The VC Control Table is the central hub of ViChaR's operation, keeping track of all in-use VCs and a detailed status of the unified buffer (UBS). When flits arrive and/or depart, the Arriving/Departing Flit Pointers Logic controls the UBS's MUXes and DEMUXes in accordance with the VC Control Table.

It is important to realize that the UBS is physically identical to the generic buffer structure: the v independent k-flit FIFO buffers of a traditional implementation are simply *logically* grouped in a single vk-flit entity (the UBS in Fig. 3.4). Hence, other than the control logic, there is no additional hardware complexity, since the vk-flit UBS is NOT a large, monolithic structure; it groups the existing buffers together, and it is only through the use of its control mechanism (the UCL) that the buffers appear as a *logically unified structure*. As shown in Fig. 3.4, UBS retains the same number of MUXes/DEMUXes as the generic implementation, i.e., one MUX/DEMUX per k flits, to avoid large (and hence slower) components.

3.3.1 Variable Number of Virtual Channels

Whereas a conventional NoC router can support only a fixed, statically assigned number of VCs per input port (namely v, as shown in Fig. 3.4), the ViChaR architecture can have a variable number of assigned VCs, based on network conditions. *ViChaR assigns at most one packet to each VC* so as to enable fine flow control granularity; on the contrary, the sharing of a single VC by multiple packets can lead to situations where a blocked packet impedes the progress of another packet which happens to use the same VC (known as HoL blocking, as described in Section 3.3). A vk-flit ViChaR structure can support anywhere between v VCs (when each VC occupies the maximum of k flits) and vk VCs (when each VC occupies the minimum of 1 flit) at any given time under full load. This variability in the number of in-use VCs is illustrated in Fig. 3.5. To aid understanding, each VC in Fig. 3.5 is shown to occupy a contiguous space in the buffer (UBS); in reality, however, this may not be the case because the UBS allows the VCs to include non-consecutive buffer slots (this fact will be explained in more detail in Section 3.3.2). Hence, the

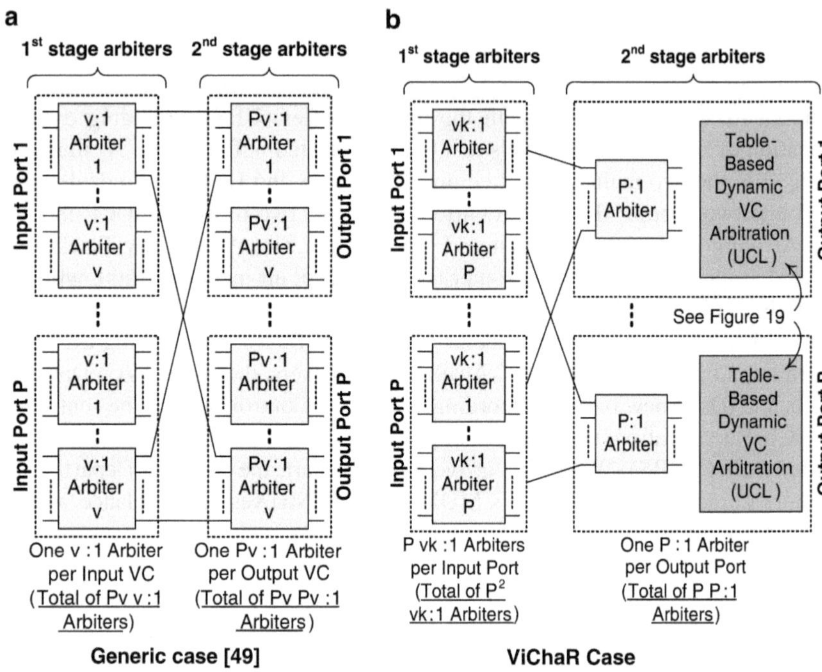

Fig. 3.7 Virtual channel arbitration (VA). (**a**) Generic case [49]. (**b**) ViChaR case

system can support a variable number of in-flight packets per port, dynamically allocating new VCs when network conditions dictate it. Dynamic variability in in-flight messages can increase throughput under heavy traffic.

As a result of its unified buffer and dynamic behavior, the ViChaR structure alters the Virtual channel Allocation (VA) logic of the router. Since the router function may return multiple output VCs restricted to a single physical channel [49], two arbitration stages are required in both the generic and ViChaR cases, as shown in Fig. 3.7. In the generic case, the first stage reduces the number of requests from each input VC to one (this ensures the request of a single VC at a particular output port by each input VC). Subsequently, the winning request from each input VC proceeds to the second arbitration stage. Details of the VA operation are omitted for brevity, but can be found in [49].

In the proposed ViChaR architecture, VA takes a different approach due to the dynamic VC allocation scheme: the first arbitration stage reduces the number of requests for *a particular output port* to one request per input port. The generic router (Fig. 3.7a) requires v:1 arbiters, since the number of VCs supported is fixed to v. ViChaR, on the other hand, supports anywhere between v and vk VCs per port at any given time. To accommodate the worst case scenario (i.e., vk in-flight VCs), ViChaR needs larger vk:1 arbiters in stage 1 of the allocation (Fig. 3.7b). The second arbitration stage in ViChaR produces *a winner for each output port among all the competing input ports*. Therefore, while the proposed ViChaR architecture uses

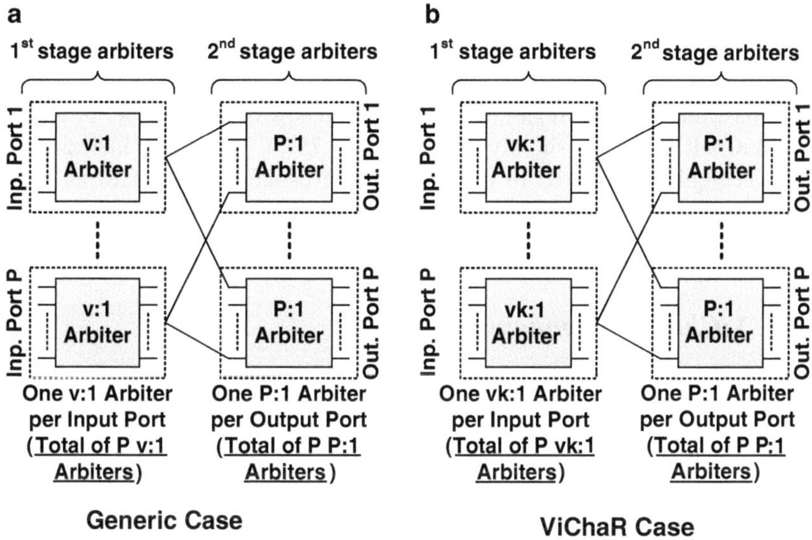

Fig. 3.8 Switch allocation (SA). (**a**) Generic case. (**b**) ViChaR case

larger Stage 1 arbiters (*vk*:1 vs. *v*:1), it uses much smaller and fewer Stage 2 arbiters. The reason for the simplified second stage is that ViChaR dynamically allocates VCs as needed, instead of accepting requests for specific VCs (which would necessitate one arbiter per output VC, just like the generic case). It is this attribute that helps the ViChaR implementation incur only a slight increase in power consumption (and even achieve a small area decrease), compared to a generic architecture, as will be shown shortly.

The variable number of VCs supported by ViChaR also necessitates bigger arbiters in the first stage of Switch Allocation (SA), as shown in Fig. 3.8. Similar to VA, switch allocation is performed in two stages. The first stage accounts for the sharing of a single port by a number of VCs. Again, ViChaR needs larger *vk*:1 arbiters. The second stage arbitrates between the winning requests from each input port (i.e., *P* ports) for each output port; thus, it is the same for both architectures. The ViChaR overhead due to the bigger stage-1 SA arbiters (illustrated in Table 1's detailed breakdown) is almost fully amortized by the bigger savings resulting from the smaller VA stage discussed previously.

To analyze the area and power overhead, NoC routers with (a) a generic buffer and (b) the proposed ViChaR buffer were implemented in structural Register-Transfer Level (RTL) Verilog and then synthesized in Synopsys Design Compiler using a TSMC 90 nm standard cell library. The resulting designs operate at a supply voltage of 1 V and a clock frequency of 500 MHz. The routers have 5 input ports (i.e., $P=5$), 4 VCs per input port (i.e., $v=4$), each VC is four-flit deep (i.e., $k=4$), and each flit is 128 bits long. Both area and power estimates were extracted from the synthesized router implementations. A comparison of the area and power overhead of the two schemes is shown in Table 1. Note that both routers have equal buffer space ($vk = 16$ buffer slots per input port) for fairness. It is evident that while ViChaR incurs an overhead in terms of control logic and switch allocation (SA), this overhead is over-compensated (in terms of area) by a larger reduction in the VA logic. Thus, the ViChaR model provides area savings of around 4%. In terms of power, ViChaR consumes slightly more power (1.75%). This power increase, however, is negligible compared to the performance benefits of ViChaR, as will be demonstrated in Section 3.4.

3.3.2 ViChaR Component Analysis

The key challenges in designing ViChaR were to *avoid* (a) deepening the router's pipeline, and (b) decreasing the operating frequency. To circumvent the multi-cycle delay induced by a linked-list approach [61] to ViChaR, we opted instead for a *table-based* approach, as illustrated in Fig. 3.10. This logic is required for each output port in the router. Following is a break-down of the control logic (UCL) sub-modules of the proposed ViChaR architecture:

VC Control Table The VC Control Table (see Fig. 3.10) forms the core of the control logic of ViChaR. It is a compact table, holding the slot IDs of all flits currently in the buffers, which are requesting the particular output port (e.g., West). Note that since the number of buffer slots in on-chip routers is resource-constrained, the size of the table is minimal, as demonstrated by the low overhead in the control logic in Table 3.1. The VC Control Table is organized by VC ID, with each VC having room for at most a single packet. Without loss of generality, in this work we assumed a packet to consist of four flits: a Head flit, two Data (middle) flits, and a

Table 3.1 Area and power overhead of the ViChaR architecture. The results in this table assume equal-size buffers for both router designs. However, ViChaR's efficient buffer management scheme allows for a 50% decrease in buffer size with no performance degradation (see Section 3.4). In such a case, area and power is reduced by 30% and 34%, respectively, over the whole router

Component (one input port)	Area (in μm^2)	Power (in mW)
ViChaR table-based control logic	12,961.16	5.36
ViChaR buffer slots (16 slots)	54,809.44	15.36
ViChaR VA logic	27,613.54	8.82
ViChaR SA logic	6,514.90	2.06
TOTAL for ViChaR architecture	101,899.04	31.60
Generic control logic	10,379.92	5.12
Generic buffer slots (16 slots)	54,809.44	15.36
Generic VA logic	38,958.80	9.94
Generic SA logic	2,032.93	0.64
TOTAL for generic architecture	106,181.09	31.06
ViChaR overhead/savings	–4,282.05	+0.54
	4.03% savings	1.74% overhead

Tail flit. The packet size is assumed to be constant, but the table can trivially be changed to accommodate a variable-sized packet protocol. As seen in the VC Control Table box of Fig. 3.10 (right-hand side), the VCs can include non-consecutive buffer slots (e.g., VC1 comprises of slots 2, 4, 6 and 7) of the South input port (i.e., flits arriving from the South). This attribute allows full-flexibility in buffer utilization and avoids the issues encountered in statically-allocated buffers. VC3 only occupies one slot (10) in Fig. 3.10. In a static buffer, 3 additional slots would have to be reserved for the remaining flits of VC3; those slots would remain unused if the remaining flits happened to be blocked in previous routers. Instead, in ViChaR those slots can be used by other VCs, thus maximizing the buffer utilization. Furthermore, the use of a table-based controller makes the management of a variable number of VCs very easy: non-used VCs are simply NULLed out in the VC Control Table (e.g., VC4 in Fig. 3.10).

Arriving/Departing Flit Pointers Logic The Flit Pointers Logic directly controls the Input and Output MUXes/ DEMUXes of the unified buffer (UBS), as illustrated in Fig. 3.9, and is directly linked to the VC Control Table module. Once a flit departs, its location in the VC Control Table is invalidated by asserting a NULL bit. There is a set of such pointers for each VC in the table. However, the overhead is minimal due to the simplicity of the pointer logic; both Departing and Arriving Flit Pointers are implemented in combinational logic and simply have to observe the non-NULL locations in their VC. For example, the Departing Flit pointer points at the first non-NULL location (in its particular VC) starting from the left of the table, as shown on the right side of Fig. 3.10 for VC2 (in the VC Control Table box). If all the entries in a single row of the VC Control Table are NULL, then the VC must be empty; thus, the pointer logic releases the VC by notifying the VC Availability

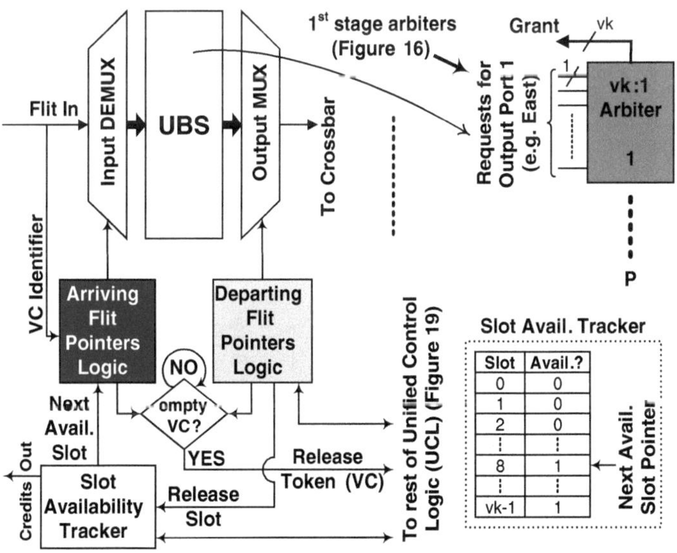

Fig. 3.9 The ViChaR UBS architecture (one input port shown)

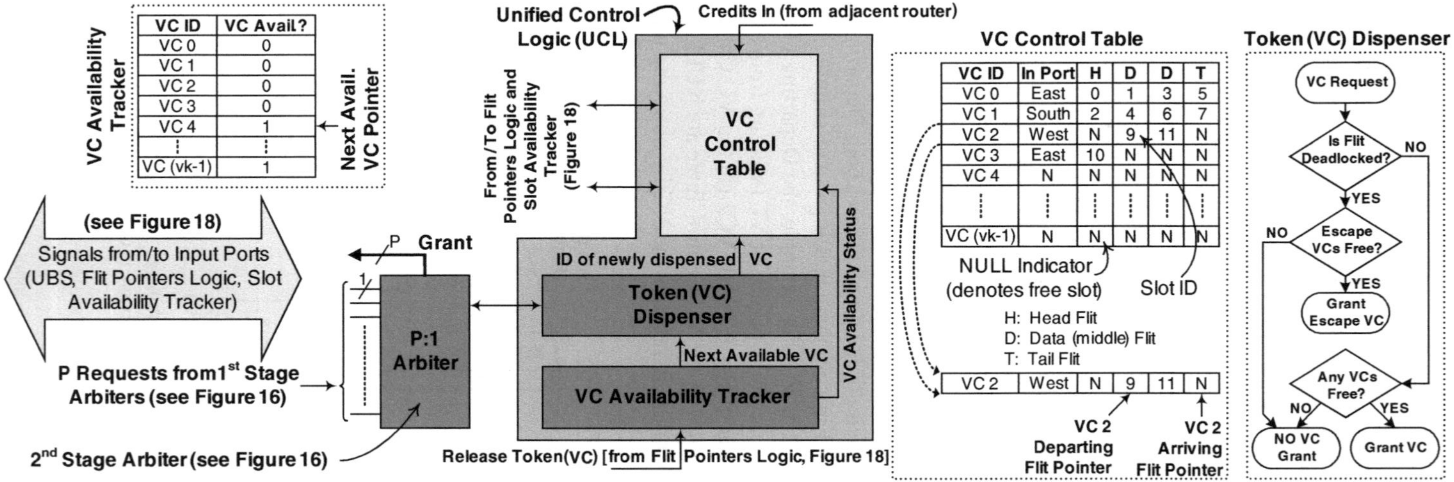

Fig. 3.10 ViChaR table-based UCL architecture (logic for one of P ports shown)

Tracker (Release Token signal in Fig. 3.9). When a new flit arrives, the pointer logic guides the flit to the appropriate slot in the unified buffer (UBS), based on the flit's VC ID and information from the Slot Availability Tracker. Finally, newly arrived header flits in the UBS can request an output VC by first undergoing local (1st stage) arbitration (top right of Fig. 3.9), and then global (2nd stage) arbitration (bottom left of Fig. 3.10).

VC and Slot Availability Trackers The VC Availability Tracker simply keeps track of all the VCs in the VC Control Table that are not used. The Token (VC) Dispenser dynamically assigns VCs to new incoming packets based on information provided by the VC Availability Tracker. Similarly, the Slot Availability Tracker keeps track of all the UBS slots that are not in use. When a new flit arrives, it is stored into a slot indicated by the Slot Availability Tracker. The VC and Slot Availability Trackers are functionally identical. They consist of a small table, as shown at the bottom right of Fig. 3.9 (Slot Availability Tracker) and the top left of Fig. 3.10 (VC Availability Tracker). Each row of the table corresponds to one VC ID (in the VC Availability Tracker) or one buffer slot (in the Slot Availability Tracker). For each entry in the table, one bit indicates that the VC/Slot is available (logic 1) or occupied (logic 0). Both trackers have a pointer that points to the top-most available entry. If all VCs are occupied (i.e., all-zero table in the VC Availability Tracker), the Token (VC) Dispenser stops granting new VCs to requesting packets. Similarly, an all-zero table in the Slot Availability Tracker implies a full buffer (UBS); this is reflected in the credit information sent to adjacent routers. The functionality of the VC/Slot Availability Trackers is implemented in combinational logic, similar to the Flit Pointers Logic described above.

Token (VC) Dispenser The Token (VC) Dispenser interfaces with the P:1 Arbiter and is responsible for dispensing free VCs to requesting packets. VCs here are like tokens; they are granted to new packets and then returned to the dispenser upon release. The flow diagram of the Dispenser's operation is illustrated on the right-hand side of Fig. 3.10. Based on information provided by the VC Availability Tracker, the Token Dispenser decides whether to grant a VC or not. The VC dispenser keeps checking for possible deadlock situations among the in-use VCs. Deadlocks may occur in networks which employ adaptive routing schemes. If a pre-specified time threshold is exceeded, the Token Dispenser can channel an existing VC into one of the escape VCs to break the deadlock. As a proof of concept of ViChaR's functionality, the experiments in this report use deterministic (XY) routing, which is inherently deadlock-free. However, ViChaR was designed to operate under adaptive routing schemes as well. Therefore, the Token (VC) Dispenser needs to account for possible deadlock situations. Toward that extent, a number of VCs can be designated as "escape", or "drain" channels to provide deadlock recovery in adaptive routing algorithms (escape channels employ a deterministic routing algorithm to break the deadlock) [71]. The Dispenser needs to switch deadlocked flits into these escape channels, if there is a need. One experiment in Section 3.4.2 validates the effectiveness of this technique under adaptive routing.

Assuming that no deadlock situation exists, the Token Dispenser can proceed with its normal operation. The Dispenser grants new VCs on a First-Come-First-Served (FCFS) basis; if a new header flit wins the VC arbitration and the VC Availability Tracker indicates that a free VC is available, then the new packet will be granted a new VC. The Dispenser does not give priority to flits of existing VCs. In principle, a more elaborate mechanism could be used for dispensing new VCs, which would monitor on-going traffic and reach a decision based on some quantitative metric or prior traffic history. However, given the highly restrictive objective function of minimal area, power and latency budgets in the design of on-chip networks, such complex monitoring mechanisms were deemed infeasible. After all, ViChaR was architected to operate within one clock cycle. The use of an FCFS scheme in the Token Dispenser turns out to be very efficient at maximizing performance. The Dispenser is able to self-throttle the dispensing of new VCs based on traffic conditions: if more packets request a channel (high traffic) more VCs are dispensed; if fewer packets are present (low traffic) fewer VCs are granted and more buffer depth is allotted to existing VCs.

ViChaR's Effect on the Router Pipeline The control logic (UCL) of ViChaR was designed in such a way as to decouple the operation of the sub-modules from each other. Thus, sub-modules are kept compact and can all operate in parallel, hence completing the entire operation in a single clock cycle. This is a significant improvement over the three-clock cycle delay of [61]. Figure 3.11 shows the pipeline stages of both a generic and the ViChaR router pipelines. As previously mentioned, the ViChaR architecture modifies the VA and SA stages (stages 2 and 3 in Fig. 3.11). The dark-colored boxes indicate the components modified/added in the ViChaR structure as compared to the generic case. As shown in the figure, the additional hardware operates in parallel without affecting the critical path of the router. This fact is also verified by our studies of the critical path delays of all

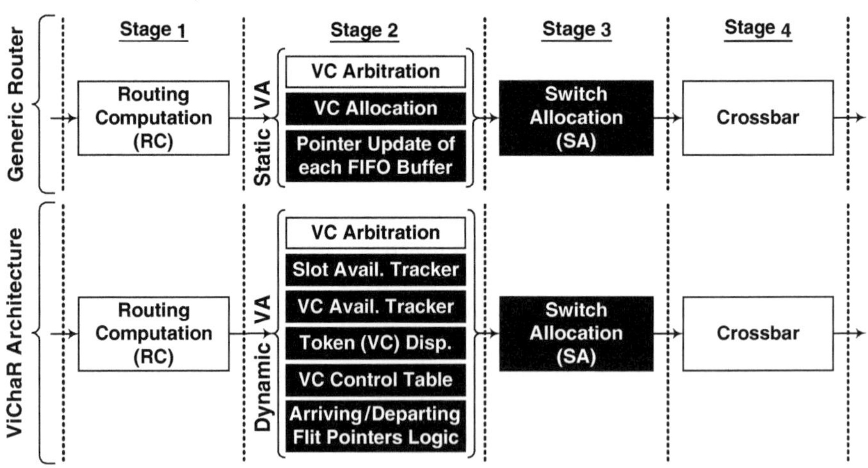

Fig. 3.11 Generic and ViChaR NoC router pipelines

major components of the router architectures (extracted from the synthesized designs). In both cases, the bottleneck that determines the minimum clock period is the arbitration logic (for the VA and SA stages, as shown in Figs. 3.7 and 3.8, respectively). All the components of the ViChaR router remain within the slack provided by the slower arbiters. Hence, the ViChaR architecture does not affect the pipeline depth or the clock frequency. Furthermore, since ViChaR does not create any interdependencies between pipeline stages, it can also be used in speculative router architectures which minimize the pipeline length.

3.4 Simulation Results

3.4.1 Simulation Platform

A cycle-accurate on-chip network simulator was used to conduct detailed evaluation of the architectures under discussion. The simulator operates at the granularity of individual architectural components. The simulation test-bench models the pipelined routers and the interconnection links. All simulations were performed in a 64-node (8×8) MESH network with 4-stage pipelined routers. Each router has 5 physical channels (ports) including the PE-router channel. The generic router (shown as "GEN" in results graphs) has a set of 4 virtual channels per port. Each VC holds four 128-bit flits (i.e., a total of $5 \times 4 \times 4 = 80$ buffer slots). The ViChaR router ("ViC" in results graphs) has a 16-flit unified buffer per port (i.e., a total of $5 \times 16 = 80$ buffer slots, just like the generic case). One packet consists of four flits. The simulator keeps injecting messages into the network until 300,000 messages (including 100,000 warm-up messages) are ejected. Two network traffic patterns were investigated: (1) Uniform Random (UR), where a node injects messages into the network at regular intervals specified by the injection rate, and (2) Self-Similar (SS), which emulates internet and Ethernet traffic. For destination node selection, two distributions were used: (1) Normal Random (NR), and (2) Tornado (TN) [72]. In all cases, except one, deterministic (XY) routing and wormhole switching were employed. One experiment used minimal adaptive routing to evaluate the systems in a deadlock-prone environment. Single link traversal was assumed to complete within one clock cycle at 500 MHz clock frequency. Both dynamic and leakage power estimates were extracted from the synthesized router designs and back-annotated into the network simulator.

3.4.2 Analysis of Results

Our simulation exploration starts with a latency comparison between a conventional, statically assigned buffer architecture and the proposed ViChaR implementation.

We first assume that both designs have equal-sized buffers; specifically, 16-flit buffers per input port (i.e., a total of 80 flits per NoC router).

In the generic design (GEN), the 16 buffer slots are arranged as 4 VCs, each with a 4-flit depth. ViChaR (ViC), on the other hand, can dynamically assign its 16 buffer slots to a variable number of VCs, each with a variable buffer depth. Figure 3.12a, b show the average network latency (in clock cycles) as a function of injection rate (in flits/node/cycle) for Uniform Random (UR) and Self-Similar (SS) traffic patterns, respectively. The graphs include results for both Normal Random (NR) and Tornado (TN) source-destination selection patterns. In all cases, ViChaR substantially outperforms the generic architecture; by 28% (NR) and 24% (TN) on average for Uniform Random traffic, and 25% (NR) and 18% (TN) for Self-Similar traffic. More importantly, though, ViChaR saturates at higher injection rates than the generic case.

Figure 3.12c shows the buffer occupancy at injection rates between 0.25 and 0.35 (i.e., before the onset of saturation). Higher buffer occupancy indicates network blocking. ViChaR is clearly much more efficient at moving flits through the router; the buffer occupancy of a 16-flit/port ViChaR design is considerably lower than an equal-size static configuration. *Buffer occupancy alone, however, is not an indicative metric, since it does not relay any information about network latency.* To validate ViChaR's highly efficient buffer management scheme, its latency at these smaller buffer sizes should also be investigated. To that extent, Fig. 3.12d and Fig. 3.12e depict how the latency of ViChaR at various buffer sizes compares to the latency of the generic architecture with a fixed 16-flit/port buffer size. It is evident from the graphs that *the UBS can achieve similar performance with less than half the buffer size of the generic architecture*. This is of profound importance, since buffers dominate the area and power budgets of NoC routers; reducing the buffer size by 50% will yield significant savings. An alternative way to visualize this phenomenon is illustrated in Fig. 3.12f. This graph shows how the latency of ViChaR at various buffer sizes compares to the latency of the generic architecture with a 16-flit/port buffer size (horizontal dashed line) at an injection rate of 0.25. ViChaR has higher latency only when its buffer size drops below 8 flits per port. On the other hand, Fig. 3.12g shows that decreasing the buffer size in a generic, statically assigned buffer structure always degrades performance.

Following on the very encouraging result that ViChaR can achieve similar performance as a conventional buffer by using only half the buffers, Fig. 3.12h shows the total average power consumption of the 8×8 MESH network for different buffer configurations. For equally sized configurations, ViChaR consumes slightly more power than a conventional buffer structure. At injection rates up to 0.3, ViChaR consumes about 2% more power, corroborating the results of Table 1. At higher injection rates (when the network saturates), excessive switching activity causes this difference to grow a bit more, even though it never exceeds 5%. However, since ViChaR's efficiency allows us to *halve* the buffer resources with no discernible effect on performance, the overall power drops by about 34% (ViC-8 in Fig. 3.12h) for equivalent performance. Similarly, the area occupied by the router decreases by around 30%, based on synthesis results. These decreases can lead to more power- and area-efficient SoCs.

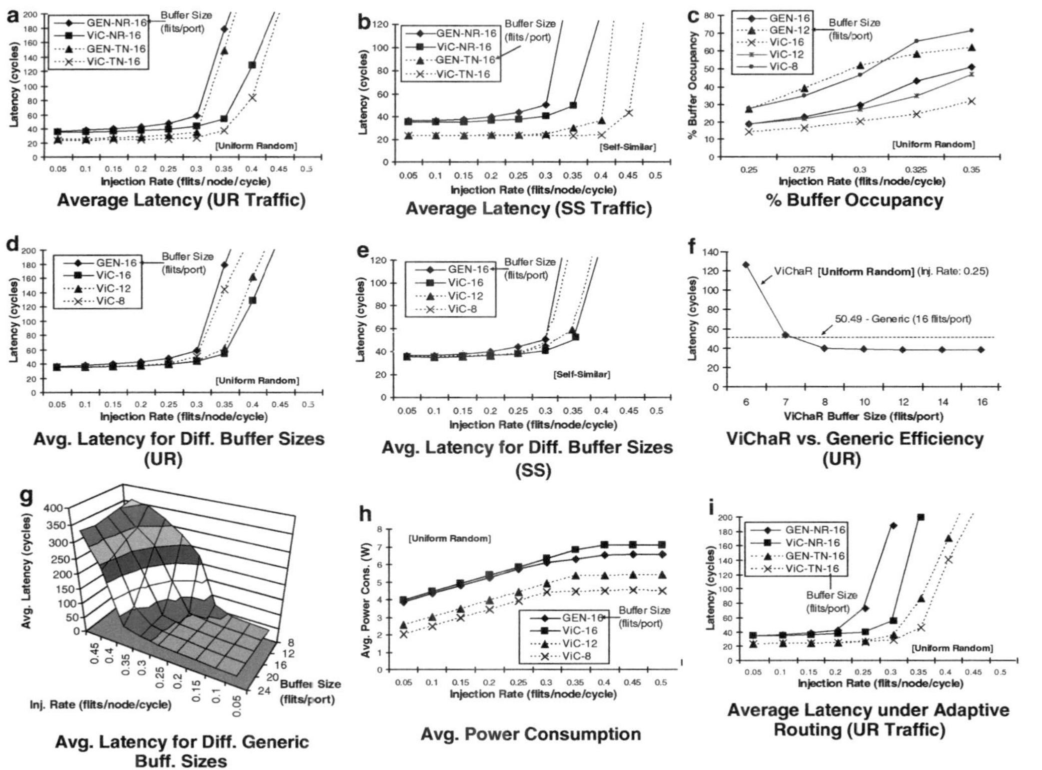

Fig. 3.12 Average latency, % buffer occupancy, and average power consumption simulation results. (**a**) Average latency (UR traffic). (**b**) Average latency (SS traffic). (**c**) % Buffer occupancy. (d) Average latency for different buffer sizes (UR). (**e**) Average latency for different buffer sizes (SS). (**f**) ViChaR vs. generic efficiency (UR). (**g**) Average latency for different generic buffer sizes. (**h**) Average power consumption. (**i**) Average latency under adaptive routing (UR traffic)

Figure 3.12i compares average network latency under minimal adaptive routing to validate ViChaR's effectiveness in handling deadlocks. Escape (drain) channels, which employ deterministic (i.e., deadlock-free) routing, were used in both the generic and ViChaR architectures to break deadlocks. Evidently, ViChaR was able to handle all deadlock situations while significantly outperforming the conventional design.

Figure 3.13a, b present another metric of network performance, namely, throughput (in flits per cycle). These graphs follow the same trend as the latency experiments, with ViChaR clearly outperforming a conventional buffer structure.

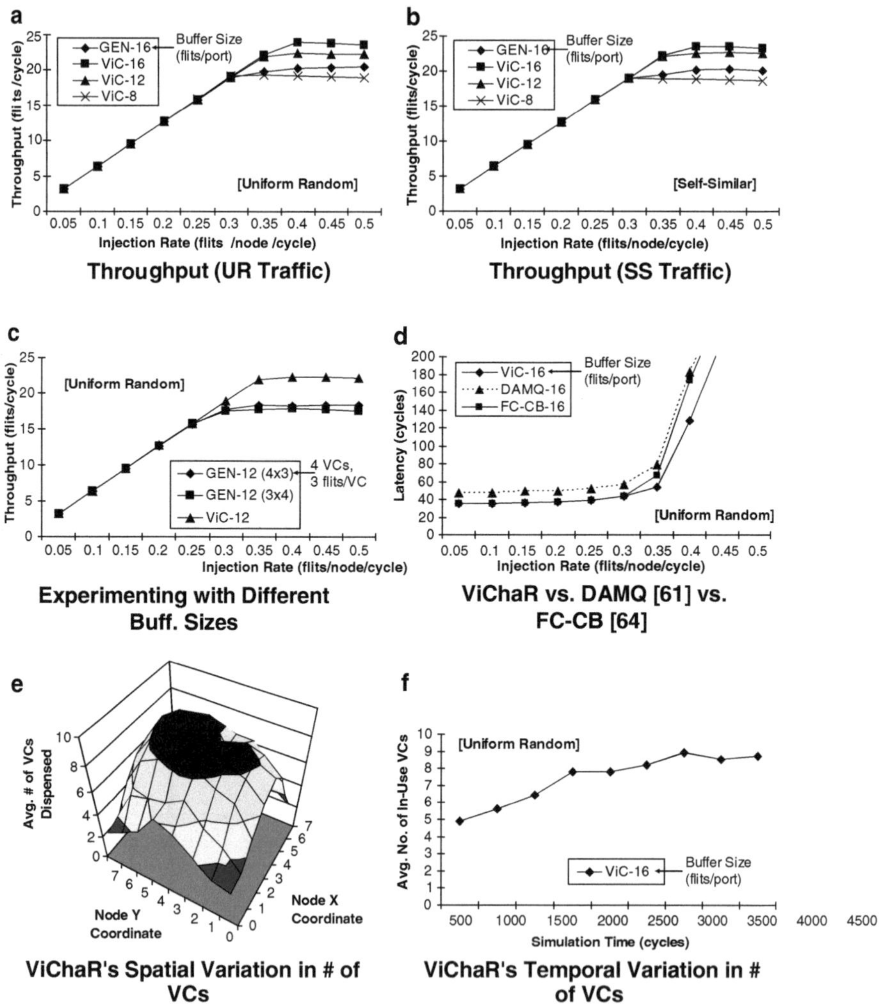

Fig. 3.13 Simulation results demonstrating ViChaR's efficient virtual channel management scheme. (**a**) Throughput (UR traffic). (**b**) Throughput (SS traffic). (**c**) Experimenting with different buffer sizes. (**d**) ViChaR vs. DAMQ [61] vs. FC-CB [64]. (**e**) ViChaR's spatial variation in # of VCs. (**f**) ViChaR's temporal variation in VCs

Figure 3.13c includes the throughput of two different (but of equal size) generic configurations: 4 VCs each with a 3-flit depth, and 3 VCs with a 4-flit depth. The graph indicates that while varying the statically-assigned VC configuration of a generic buffer does affect throughput, its performance still trails that of the dynamically variable design of ViChaR.

In the related work sub-section (Section 3.2), we analyzed in detail why the unified buffers of the DAMQ [61] and FC-CB [64] would underperform compared to ViChaR's dynamic design. Both the DAMQ and FC-CB structures were implemented and incorporated into our cycle-accurate simulator. Figure 3.13d shows how all designs fare against each other. DAMQ loses out because of its 3-cycle buffer delay, and its fixed number of VCs, as previously explained. For a fair comparison, we assumed that the FC-CB design completes its buffer management procedure in one clock cycle (just like ViChaR). As seen in Figure 3.13d, at low injection rates, the FC-CB's performance is almost identical to ViChaR's. However, as network traffic increases, FC-CB's performance starts to degrade compared to ViChaR. This is attributed to FC-CB's fixed number of VCs (i.e., just like DAMQ). Under heavier traffic loads, ViChaR's ability to dynamically dispense more VCs helps improve performance quite drastically. Note also that both FC-CB and DAMQ would incur much higher area and power penalties (as explained in Section 3.2). In terms of throughput (not shown here), ViChaR's improvement over DAMQ and FC-CB is a more modest 5% (on average). However, ViChaR would perform substantially better in latency-critical applications.

Finally, Figure 3.13e depicts the spatial variation in the number of VCs used in the 8×8 MESH, while Figure 3.13f shows the temporal variation over simulation time. The average number of VCs used varies continuously according to network traffic. Figure 3.13e shows the average number of VCs dispensed at each node of the 8×8 MESH network over the whole simulation time at an injection rate of 0.25. As expected, the nodes situated at the middle of the network exhibit higher congestion; ViChaR successfully self-throttled its resources by granting more VCs in these nodes in order to optimize performance.

Figure 3.13f, as the network fills up with packets, the average number of VCs used over all nodes increases accordingly to handle the traffic. These results validate the effectiveness of our FCFS scheme employed in the Token (VC) Dispenser (Section 3.3.2).

3.5 Chapter Summary

NoC performance is directly related to the routers' buffer size and utilization. In this chapter, the author introduced a centralized buffer architecture, called the Virtual Channel Regulator (ViChaR), which dynamically allocates virtual channels and buffer slots in real-time, depending on traffic conditions. Unlike current implementations, the ViChaR can dispense a variable number of VCs at any given time to maximize network throughput.

Simulation results using a cycle-accurate network simulator indicate *performance improvement of around 25%* under various traffic patterns, as compared to a conventional router with equal buffer size, with a modest 2% power increase. Most importantly, though, ViChaR is shown to achieve performance similar to that of a generic router, while using a 50% smaller buffer. This attribute is a testament to ViChaR's efficient dynamic buffer management scheme, and is a result of utmost significance in the NoC arena. Synthesized designs in 90 nm technology indicate that *decreasing the ViChaR's buffer size by 50% leads to area and power savings of 30% and 34%, respectively, with no degradation in performance.*

Chapter 4
RoCo: The Row–Column Decoupled Router – A Gracefully Degrading and Energy-Efficient Modular Router Architecture for On-Chip Networks [40]

Network-on-chip architectures are required to not only provide ultra-low latency, but also occupy a small footprint and consume as little energy as possible. Further, reliability is rapidly becoming a major challenge in deep sub-micron technologies due to the increased prominence of permanent faults resulting from accelerated aging effects and manufacturing/testing challenges.

Towards the goal of designing low-latency, energy-efficient and reliable on-chip communication networks, we propose a novel fine-grained modular router architecture. While Chapter 3 focused solely on the router buffers, this chapter will attempt to optimize several of the remaining core micro-architectural components. In particular, the proposed architecture employs decoupled parallel arbiters and uses smaller crossbars for row and column connections to reduce output port contention probabilities as compared to existing designs. Furthermore, the router employs a new switch allocation technique known as "Mirroring Effect" to reduce arbitration depth and increase concurrency. In addition, the modular design permits graceful degradation of the network in the event of permanent faults and also helps to reduce the dynamic power consumption. Our simulation results indicate that in an 8×8 mesh network, the proposed architecture reduces packet latency by 4–40% and power consumption by 6–20% as compared to two existing router architectures. Evaluation using a combined performance, energy and fault-tolerance metric indicates that the proposed architecture provides 35–50% overall improvement compared to the two earlier routers.

4.1 Introduction and Motivation

While researchers have proposed performance enhancement techniques [48,73], area-constrained design alternates [19,74], power-efficient and thermal-aware systems [22,70,75], and fault-tolerant mechanisms [35] for NoCs, a systematic design methodology encompassing the interplay of performance, fault-tolerance and energy constraints is yet to evolve. In an attempt to address this issue, this chapter

C. Nicopoulos et al., *Network-on-Chip Architectures: A Holistic Design Exploration*, Lecture Notes in Electrical Engineering 45,
DOI 10.1007/978-90-481-3031-3_4, © Springer Science+Business Media B.V. 2009

presents the design of a modular wormhole-switched router architecture considering the performance, energy, and fault-tolerance issues in a cohesive manner. The salient features of the proposed router that make it distinct compared to other contemporary designs are the following:

(1) For enhancing performance, the virtual channel allocation (VA) and switch allocation (SA) units are cleverly exploited in minimizing the delay due to resource contention.
(2) For energy conservation, the author focuses on use of smaller crossbars, simplified arbiter circuits, and other design tricks such as early ejection and mirrored arbitration.
(3) For enhancing fault-tolerance, the architecture relies on a modular design such that failure of a router component can be tolerated by allowing the switch to operate in a degraded mode. In this context, several techniques are proposed to handle hardware faults in different components of the router.

The proposed Row–Column (RoCo) Decoupled Router enhances performance by reducing the contention probability. This is achieved by splitting the router operation into two distinct and independent modules. Each module is responsible for handling traffic in one dimension (X-dimension or Y-dimension). This decoupling permits the use of smaller and simpler components with reduced logic depth. Each module requires a compact 2×2 crossbar, as opposed to the bigger monolithic crossbar, used in conventional architectures. Furthermore, the proposed router uses a novel switch arbitration scheme, known as the Mirroring Effect. This mechanism requires fewer global arbiters and maximizes crossbar utilization by providing optimal matching between inputs and outputs. Finally, contention in the crossbar is further reduced through the use of a preliminary path-sensitive buffering process, known as Guided Flit Queuing.

The RoCo router possesses inherent fault-tolerant attributes. Its decoupled operation allows for partial functionality in the event of a hard failure. Having two operationally independent modules implies that one module can continue to provide service in one dimension even if the other module is blocked due to a permanent failure. This alleviates contention around the faulty node, which has a profound effect on network latency. Additionally, the proposed architecture employs a hardware recycling mechanism which uses resource sharing to circumvent hard failures in various intra-router components. A comprehensive router fault model is presented and several safeguards are proposed to protect against various types of intra-router faults. These measures induce minimal area, power and latency overheads.

A flit-level, cycle-accurate simulator is used to analyze the performance of the proposed architecture and compare it against a generic 2-stage router and the Path-Sensitive router [73] under a variety of traffic patterns. Moreover, the routers are synthesized in 90 nm technology to extract their power profiles. The power numbers are imported into the simulator for detailed energy analysis. The three router types are used to analyze the performance of an 8×8 mesh network with deterministic, XY–YX and adaptive routing. Simulation results show that the proposed architecture reduces packet latency by 4–40% and power consumption by 6–20% as compared to the two aforementioned router architectures. Evaluation using a combined performance, energy and fault-tolerance metric indicates that our architecture provides a substantially improved combination of high performance, low energy and effective fault-tolerance (almost 50% improvement compared to the generic router and 35% improvement with respect to the Path-Sensitive router).

4.2 Related Work in Partitioned Router Architectures

A comprehensive survey of NoC architectures can be found in [76,77]. In this section, we concentrate predominantly on prior work in partitioned router implementations, where the router architecture is characterized by a modular or hierarchical structure that divides labor to the various components based on an underlying fundamental premise.

Kim et al. [78] proposed a hierarchical switch and cross-point buffering considering the significance of effective allocation for interconnection networks, suitable for parallel computer architectures. The design employs a hierarchical crossbar, in which the sub-switches are interconnected and require global control logic for coordination. Further, a flit may require concatenated crossbar traversal, and use VC buffers at intermediate switching points. The same philosophy is used in this chapter, i.e., *lowering contention*, but with a different design approach for NoCs. The author's approach splits the crossbar into two totally independent and decentralized switches. There is no concatenated switch traversal and no centralized control logic.

The *Path-Sensitive router* supporting routing adaptivity [73] utilizes look-ahead routing in selecting the next route. The router is called Path-Sensitive because – based on the destination address – it has four sets of VCs, called path sets; one set for possible traversal in each of the four quadrants: NE, SE, NW, and SW. Each path set has three groups of VCs to hold flits from possible directions from the previous router. The architecture utilizes a 4×4 decomposed crossbar design with half the connections of a full crossbar. It was shown that the Path-Sensitive router can reduce the average latency compared to a 2-stage router. While this is a nice approach to reduce network latency, it will be shown in this chapter that one can do better by reducing the crossbar size and employing a decoupled design for better concurrency and fault-tolerance. The Partitioned Dimension-Order Router (PDR) [79,80] uses two 3×3 crossbars. However, the operation of the two crossbars is intertwined and the flits should take concatenated switch traversals in order to change dimension. Choi and Pinkston [81] also proposed partitioned crossbar architectures exploring spatial locality and the fact that a packet tends to traverse the network in the same virtual channel. However, these schemes are different compared to the proposed architecture.

4.3 The Proposed Row–Column (RoCo) Decoupled Router

4.3.1 Row–Column Switch

Figure 4.1a illustrates the architecture of a generic 5-port, 2-stage NoC router employing virtual channel flow control and wormhole switching. As already explained in Chapter 2, the five ports correspond to the four cardinal directions and the connection to the local Processing Element (PE). The router consists of six major components: the Routing Computation unit (RC), the Virtual Channel Arbitration (VA), the Switch Allocator (SA), the MUXes and DEMUXes which control the flit flow through the router, the VC buffers, and the crossbar. It employs a pipelined design with speculative path selection to improve performance. Instead of relying on a unified architecture with a monolithic crossbar, the proposed router consists of dual compact crossbars arranged in Row and Column Path Sets. Figure 4.1b depicts the major components of the new 2-stage, pipelined router architecture. The first stage is responsible for look-ahead routing, virtual channel allocation (VA) and speculative switch allocation (SA); all three operations are performed in parallel. The second stage is responsible for crossbar traversal. In this work, the functionality of the router is described with respect to a 2D mesh interconnect.

The router has two sets of crossbars, called Row-Module (East–West) and Column-Module (North–South). The router is divided into two distinct, independent units, each responsible for possible traversal in the corresponding crossbar connections: i.e., in the East–West direction or in the North–South direction. Each port of the crossbar module has a set of three VCs to hold arriving flits from neighboring routers or the local PE. These sets are aptly named Path Sets, since all flits within such a set travel in the same physical direction. In order for an incoming header flit to pass through the DEMUX and be placed into the buffer corresponding to its output path, the header flit should know its route before departing the previous node. To remove the routing process from the router's critical path, the Routing Computation (RC) can be performed one step ahead. By employing this Look-Ahead Routing scheme, the flit is guided to the appropriate buffer by the DEMUX.

Based on the required output port, the header flit requests a valid output VC. The virtual allocation unit, VA, arbitrates between all packets requesting access to the same VCs and decides on winners. Figure 4.2 compares the complexity of the VA unit of a generic 5-port (North, East, South, West, and PE) router and the proposed RoCo router. The use of early ejection allows the RoCo router to eliminate the PE path set (early ejection is analyzed later on in this sub-section). In this comparison, we assume v VCs per input port for both the generic and RoCo architectures. Figure 4.2 compares two cases: one, where the routing function returns a single virtual channel ($R{=}{>}v$), and one, where the routing function returns a single physical channel ($R{=}{>}p$). Clearly, the RoCo router requires fewer ($4v$ vs $5v$) and smaller ($2v$:1 vs $5v$:1) arbiters in both cases. This attribute significantly reduces the complexity of the arbitration process, since smaller and fewer arbiters imply less contention

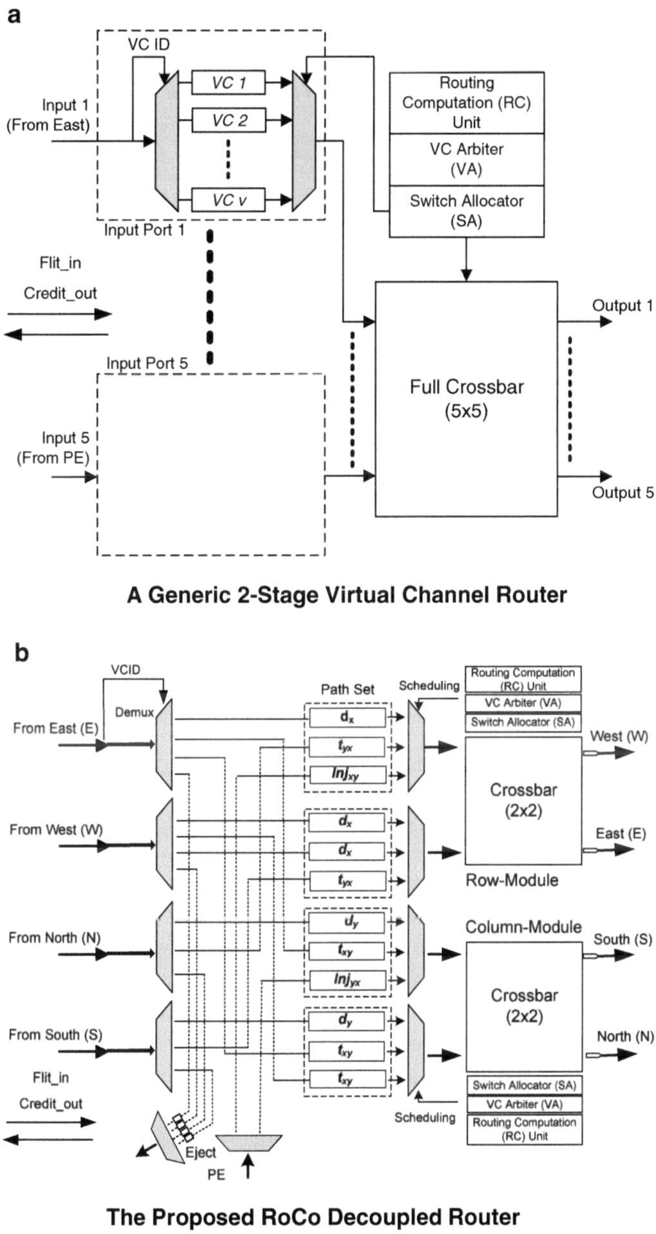

Fig. 4.1 On-chip router architectures: (**a**) a generic 2-stage virtual channel router, (**b**) the proposed RoCo Decoupled Router

and reduced arbitration depth. On the contrary, the increased complexity of the VA in a generic architecture requires multiple iterative arbitrations before satisfying all pending requests [49].

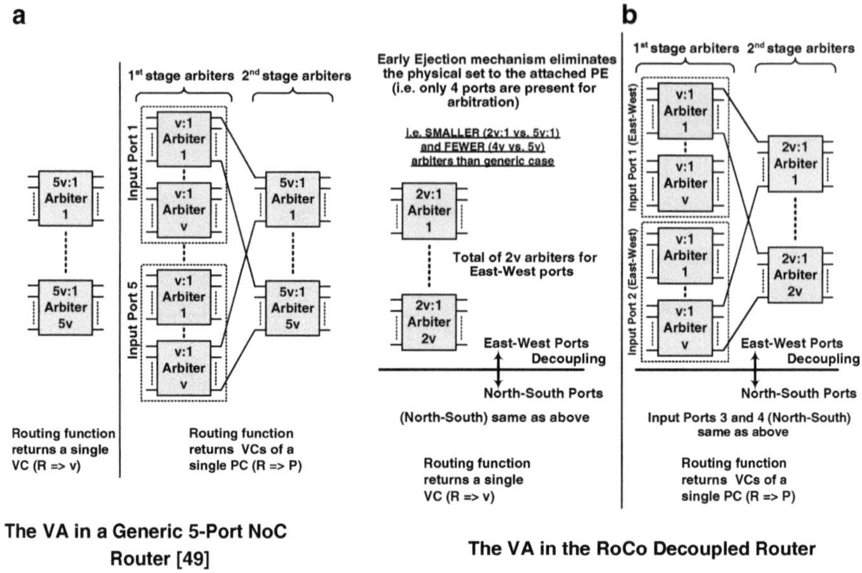

Fig. 4.2 Virtual channel arbitration (VA) comparison: (**a**) the VA in a generic 5-port NoC router [49], (**b**) the VA in the RoCo Decoupled Router

In the proposed architecture, look-ahead routing decides the valid outgoing channels of packets based on their output paths (Row-Module or Column-Module) and decides whether or not they are continuing along the same dimension. In Fig. 4.1b, VCs marked d_x (d_y) hold flits which continue traversal in their current X (Y) dimension, i.e., East or West (North or South). VCs marked t_{xy} (t_{yx}) hold flits that switch from the X to the Y dimension (Y to X). For example, a flit traversing the network from the east toward the north or the south will arrive at the t_{xy} VC of the first input port in the Column-Module. If the flit is to continue traversal to the west, it is buffered in the d_x VC. That is, d_x and d_y VCs are used for on-going flits along the same dimension, while t_{xy} and t_{yx} are used for changing from the Row-Module to the Column-Module and the other way around. A flit coming from a local PE and destined to the X-dimension, such as the east or the west outputs, is buffered in the Inj_{xy} VC of the Row-Module, while a flit addressed to the Y-dimension is queued into the Inj_{yx} VC of the Column-Module. Depending on the type of routing algorithm used in the network, the number and configuration of the VC buffers changes accordingly. A deadlock free deterministic routing algorithm, such as XY routing, requires a minimum of eight VCs for correct functionality (2 d_x, 2 d_y, 2 t_{xy}, 1 Inj_{xy} and 1 Inj_{yx} for source–destination pairs which lie in the same column). To provide support for deadlock-free XY–YX routing, one additional d_x VC and one additional d_y VC are required. Finally, to provide support for deadlock-free adaptive routing, one more t_{xy} VC and one more t_{yx} VC are needed, making it a total of 12 VCs, as shown in Fig. 4.1b.

Table 4.1 VC buffer configuration for the three routing algorithms

	Row-module		Column-module	
Input port	Port 1	Port 2	Port 1	Port 2
Adaptive	$d_x\ t_{yx}\ Inj_{xy}$	$d_x\ d_x\ t_{yx}$	$d_y\ t_{xy}\ Inj_{yx}$	$d_y\ t_{xy}\ t_{yx}$
XY–YX	$d_x\ t_{yx}\ Inj_{xy}$	$d_x\ d_x\ t_{yx}$	$d_y\ t_{xy}\ Inj_{yx}$	$d_y\ d_y\ t_{xy}$
XY	$d_x\ d_x\ Inj_{xy}$	$d_x\ d_x\ Inj_{xy}$	$d_y\ t_{xy}\ Inj_{yx}$	$d_y\ d_y\ t_{xy}$

These VCs are grouped into four path sets, each containing three VCs. When the router is used with deterministic or XY–YX routing (which can operate with less than 12 VCs), the extra VCs are re-assigned to improve performance by reducing the Head-of-Line (HoL) blocking. For example, XY routing gives rise to asymmetric utilization of the router; HoL in the X-dimension happens more frequently than in the Y-dimension, and the injection channel Inj_{xy} is much more frequently used than Inj_{yx} as a result of the routing scheme. To account for this unbalanced traffic distribution, two additional d_x VCs are assigned to the extra buffers available in the router. Similarly, all 12 VCs present in the router are assigned differently, according to the routing algorithm used. The VC buffer configurations for the three supported routing algorithms (XY, XY–YX, and adaptive) are summarized in Table 4.1.

Upon successful VC allocation and provided a buffer space is available in the downstream router, a flit requests access to the crossbar by undergoing Switch Allocation (SA). The SA arbitrates between all VCs requesting access to the crossbar and grants permission to the winning flits. The winning flits are then able to traverse the crossbar and are forwarded to the respective output links. Switch arbitration works in two stages; stage 1 requires a v-input arbiter for each input port (since there are v VCs per port). Stage 2 arbitrates between the winners from each input port and requires P P input arbiters, where P is the number of physical ports. The proposed architecture splits the SA module into two smaller modules, each responsible for a small 2×2 crossbar. The reduced number of crossbar ports minimizes the complexity of the SA modules, which function independently from each other. Operation of the SA modules is described in detail in Section 4.3.3.

Deadlock Freedom Adding extra VCs is a technique commonly used to provide deadlock freedom in adaptive routing [82,83]. The two d_x VCs in the second path set of the Row-Module provide a deadlock-free path in the East–West direction during a potential deadlock. The location of the two VCs need not be in the second path set. Interchanging the locations of the VCs in the two path sets of the Row-Module would still yield the same effect. The two t_{xy} VCs in the second path set of the Column-Module are used to ensure deadlock-free routing in case of a chained cyclic dependency. The first t_{xy} VC of the Column-Module is used for turning from the east to the south direction, and the second t_{xy} VC is used for turning from the east to the north direction. Once again, the location of these VCs may be interchanged between the two path sets of the Column-Module without affecting performance.

Early Ejection A flit destined for the local PE does not traverse the crossbar, but, instead, it is ejected immediately upon arrival (hence the "Early Ejection" mechanism). This mechanism utilizes the look-ahead routing information to detect if the incoming flit is destined for the local PE and accordingly ejects it after the DEMUX. This early ejection saves two cycles at the destination node by avoiding switch allocation and switch traversal. Also, it reduces the input load for each crossbar input port. This provides a significant advantage in terms of nearest-neighbor traffic, and can take advantage of NoC mapping which places frequently communicating PEs close to each other [84].

Modular Router and Guided Flit Queuing In our proposed architecture, the input decoders (DEMUXes) undertake a more significant role than in a generic router. In the latter, the input decoders can only distribute incoming flits to the VC buffers of a single port set (see Fig. 4.1a). In the RoCo architecture, however, the input decoders can distribute flits to multiple path sets. This mechanism amounts to a preliminary switching operation, which we call "Guided Flit Queuing", and significantly alleviates contention later on in the crossbar by pre-arranging incoming flits according to their desired output path dimension (X or Y). The area consumed by the router is dominated by the buffers. Therefore, while Guided Flit Queuing increases wiring complexity, the wires have plenty of space to be routed above the buffers in upper layers of the chip, thus imposing minimal overhead. Further, a smaller and simpler crossbar structure reduces wiring complexity.

4.3.2 Blocking Delay

Network latency consists of actual transfer time and blocking delay. The blocking delay is heavily influenced by the switch allocation strategy and the traffic pattern, while the actual transfer time is determined by the floor-plan and the topology of the network. Given that the transfer time is defined by the physical design, we address the other component of network latency, i.e., blocking delay due to contention. Contention is a result of the two arbitration processes occurring within the router: virtual channel allocation and crossbar passage (input port service scheduling and output port allocation). Figure 4.3 shows the comparison of the input contention probabilities in three different architectures (Generic, Path-Sensitive [73] and RoCo) in an 8×8 mesh network with uniform traffic pattern. The results are obtained using our cycle-accurate simulator (described in Section 4.5). In Dimension-Order Routing (DOR – XY routing), the flits of the row input are involved in more severe output conflicts than the column input, because of the nature of the routing algorithm (i.e., X first, Y next). Thus, contention at the row input is higher than at the column input, as shown in Fig. 4.3a, b. Adaptive routing is useful for avoiding local congestion, but it does not reduce the contention probability unless an efficient allocation technique is employed. In fact, adaptive routing may have poor performance with uniform traffic, as explained in [82]. It is evident from Fig. 4.3 that the generic router suffers from high contention probability, which inevitably leads to high Head-of-Line

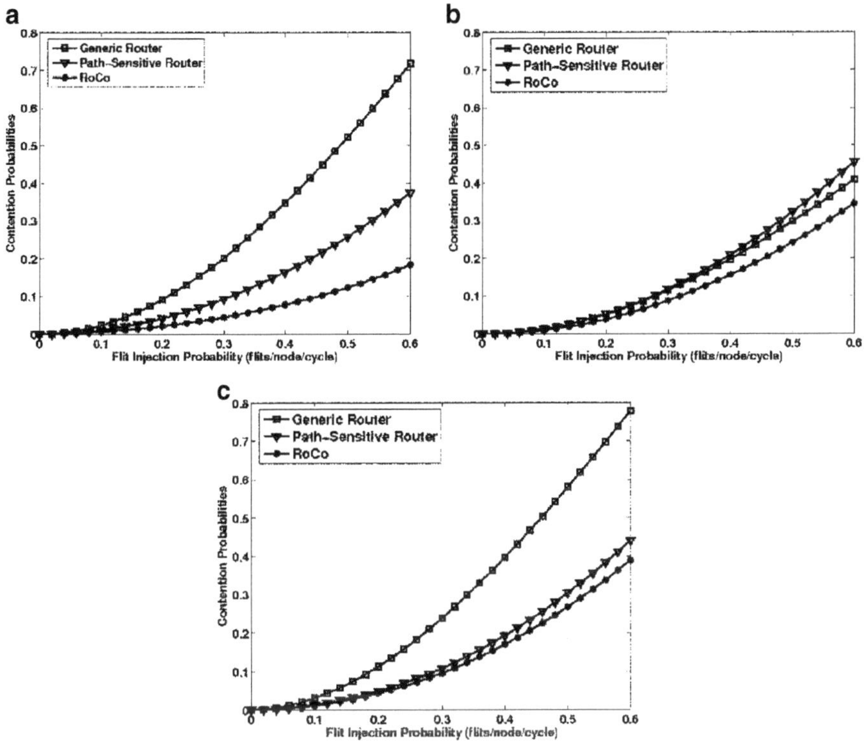

Fig. 4.3 Contention probabilities: (**a**) contention at row input in XY routing, (**b**) contention at column input in XY routing, (**c**) contention in adaptive routing

Table 4.2 Non-blocking probabilities for the three router architectures (N=5)

Route designs	Generic	Path-sensitive	Roco
Non-blocking	$0.043 = \left(\frac{F(N)}{(N-1)^N}\right)$	$0.125 = \left(\frac{2}{2^4}\right)$	$0.125 = (1-0.5)^2$

(HoL) blocking. The RoCo router has the least contention probability. Furthermore, the RoCo router significantly outperforms the other two architectures in terms of non-blocking probability (i.e., when each output port has one input connection; we call this maximal matching between input and output ports). The non-blocking probabilities for the three router architectures are shown in Table 4.2. Assuming that each input flit has an equal probability $1/(N-1)$ of accessing one of the $(N-1)$ output ports in an $N \times N$ crossbar, the number of cases in which non-blocking maximal matching, $F(N)$, occurs is computed as

$$F(N) = !-\sum_{j=1}^{N}\binom{N}{j} F(N-j) \quad where\, N >= 3, F(1) = 0\, and\, F(2) = 1 \qquad (4.1)$$

In the Path-Sensitive router proposed in [73], arriving flits are grouped in sets depending on their destination quadrant (North–East, North–West, etc.). In this architecture, two inputs from each quadrant path set request one output port. For example, flits in the two quadrants NE and NW may compete for the north output channel. In a similar fashion, two input ports also compete for one output in the RoCo router. However, RoCo uses parallel and independent crossbars, while the Path-Sensitive router has chained dependency between requests. Thus, only 2 cases out of 2^4 matches are non-blocking in the Path-Sensitive router, while 2 cases out of 2^2 matches are non-blocking in each module of the RoCo router. The RoCo router is almost six times more likely to achieve maximal matching than a generic router (25–4.3%), and two times more likely than the Path-Sensitive router (25–12.5%). This implies that the RoCo design is better in terms of providing non-blocking connections.

4.3.3 Concurrency Control for High-Contention Environments

In this section, we introduce the "Mirroring Effect", a new switching allocation scheme that provides maximal matching in the RoCo router. The Mirroring Effect is a simple algorithm that finds the maximum number of matches between inputs and outputs, customized to the small 2×2 crossbar of each module. The two sets of disjoint pair-wise switch allocators are illustrated in Fig. 4.4. The algorithm is based on the rationale that maximal matching is achieved when the switch allocation results of the two input ports of a single module are mirror images of each other. This realization allows the RoCo implementation to perform global arbitration in only one of the two input ports of each module, and the result is mirrored in the other port. This is illustrated on the right-hand side of Fig. 4.4. For example, if a flit in the top input port is to be forwarded to the West direction, then the bottom port should forward a flit to the East direction to ensure full utilization of the crossbar. Hence, the bottom input port grants access to a flit which wants to continue traversal in the East direction. This scheme constitutes a simple and concurrent global arbitration mechanism compared to the complex hierarchical arbitrations and Parallel Iterative Matching (PIM) [82]. Even though the global switch arbitration decision in the proposed Mirror Allocator is made at the first port, the allocator also gets state information from the bottom port (Fig. 4.4), ensuring that maximal matching is always achieved at each crossbar.

The proposed mechanism requires two arbiters per input port for the first (local) stage of arbitration, as opposed to just one in the generic case. The two arbiters are required to ensure maximal matching by providing the winning requests for both directions (East–West or North–South). However, this small overhead is compensated by the fact that only one arbiter is required per module (because of the Mirroring Effect) in the second (global) arbitration stage (see Fig. 4.4). The mirror arbiter is ideal for a high-throughput switch, because it resolves HoL blocking, eliminates iterative arbitrations, and reduces the inefficiency of local arbitration.

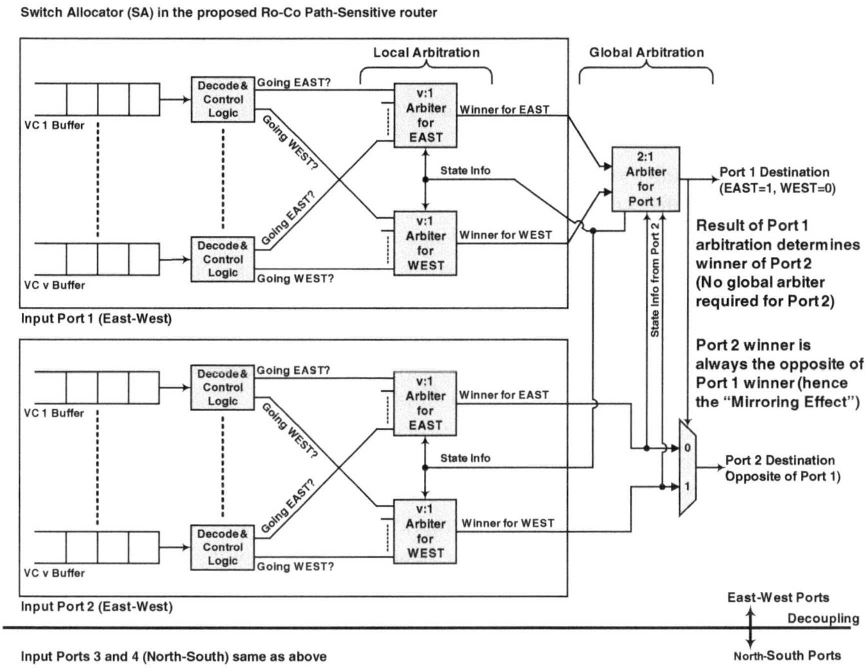

Fig. 4.4 The proposed Mirror Allocator

4.3.4 Flexible and Reusable On-Chip Communication

The routing logic, virtual channel arbitration, switch allocation and switching hardware are all partitioned into two separate and independent modules (row and column sets). This decoupling allows for partial operation in case a component within the router malfunctions or suffers a hard failure. In generic router architectures, a hard failure may cause the entire node to be taken off-line, since the operation of the router is unified between all components. In the RoCo router, however, the two disjoint modules function independently. Should a component fail, only the affected module is isolated, with full operation in the remaining module still possible. This would allow the afflicted router to handle network traffic, albeit in limited directions. The fault-tolerance advantages of the RoCo router are analyzed in Section 4.4.

4.4 Fault-Tolerance Through Hardware Recycling

Utilizing the properties of the proposed modular router, a new "Hardware Recycling" mechanism is introduced to ensure fault-tolerant operation of the router. In this section, we explore various possible failure modes within an NoC router, and propose detailed recovery schemes with minimum area and power cost. Our proposed RoCo router architecture possesses some inherent fault-tolerance due to its decoupled design. This additional operational granularity may be utilized to allow replacement of a faulty component by another one, thus allowing partial operation of the router instead of a complete breakdown. The substitution of defective elements by healthy ones elsewhere in the system provides a kind of virtual recycling bin, where functional components can be reused in other parts of the implementation should the need arise. Our proposed scheme avoids the more traditional approach in fault-tolerance, which resorts to replication of resources. Silicon real-estate and energy are at a premium in on-chip applications, thus necessitating the efficient re-use of existing resources.

The six major components of the router – the Routing Computation Unit (RC), the Virtual Channel Arbiter (VA), the Switch Allocator (SA), the MUXes and DEMUXes which control the flit flow through the virtual channel buffers, the VC buffers, and the crossbar – are susceptible to different types of permanent faults. These components can be classified into two categories, based on their operational regime: (a) per-packet components, and (b) per-flit components. Per-packet components (i.e., the RC and VA) are only used to process the header flit of a new incoming packet. The subsequent flits simply follow the wormhole created by the header flit. Per-flit components (i.e., the remaining components) are used to process every single flit passing through the router. Clearly, since the per-packet based components are driven only by the header flit, their utilizations are relatively low compared to the flit-by-flit operation of per-flit components; the latter are fully utilized in non-blocked operation. Thus, packet-based resources can be shared during their unloaded periods.

We further sub-divide the fundamental router components into two classes: message-centric and router-centric. A message-centric component requires a single individual packet as its input, and does not exhibit any interdependencies with other incoming messages. The Routing Computation Unit (RC) and the virtual channel buffers are such examples; they operate on a single message (i.e., packet) and their operation does not require state information from other components within the router. On the other hand, router-centric components require inputs from several pending messages in order to execute their function. The VA and SA are such examples; they arbitrate between all messages requesting passage through the router, and their functionality requires state information from the buffers and adjacent routers.

Finally, it is important to note that the operation of the router consists of a critical pathway and non-critical control logic. The datapath of the router (i.e., guided passage of a flit and switch traversal) constitutes the critical pathway; it consists of buffers, decoders, multiplexers and the crossbar. It should be noted that even though the VC buffers lie in the critical datapath, they may or may not be classified as critical, depending on the presence or not of a bypass path. If bypass paths are employed in the buffers for performance optimization, then the VC buffers can be classified as non-critical because of the redundancy supplied by the extra path as explained later on in the section. Otherwise, the buffers are classified as critical. The operation of the control logic – comprised mostly of the arbiters of the VA and SA – lies in a non-critical pathway. Table 4.3 illustrates the fault classifications of the router components.

Table 4.3 Component fault classification

Fault type	Per-flit operation: Critical pathways	Per-flit operation: Non-critical pathways	Per-packet operation: Critical pathways	Per-packet operation: Non-critical pathways
Message centric	MUX/DEMUX buffer (w/o bypass path)	Buffer (with bypass path)	–	RC
Router centric	Crossbar	SA	–	VA

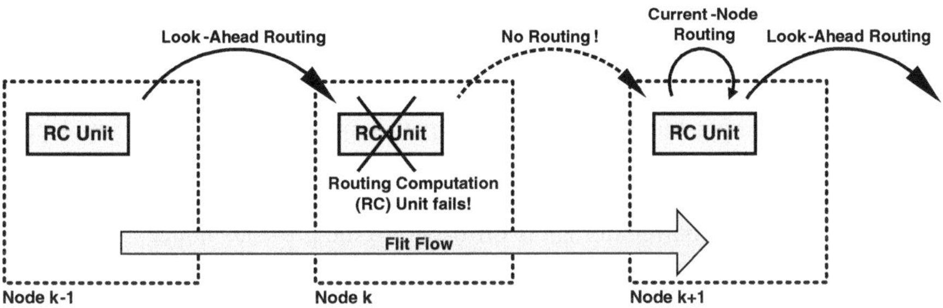

Fig. 4.5 Double Routing mechanism in the event of RC unit failure

Each router node is assumed to be able to detect a faulty component through the use of simple control signals. The novelty in our approach lies in the reaction of the router to a hard failure. If a faulty component belongs to a message-centric and non-critical region, the failure can be bypassed instead of resorting to blocking of the whole router module (Row-Module or Column-Module). We can still partially use the router module with the faulty component. If the faulty block lies on the critical pathway, or if it is a router-centric component, the permanent failure cannot be bypassed. In this case, the module is isolated and the router remains partially operational through the use of the other parallel module in our proposed scheme. Operational state is tracked by neighboring routers through the use of simple handshaking signals. We assumed permanent failures to be handled statically. Upon failure, any fragmented packets are simply discarded. Most of the fault-tolerant schemes proposed in this work can be retrofitted to existing router designs. However, they are particularly amenable to the RoCo router because of its decoupled nature that allows for graceful degradation. The recovery schemes proposed for each component failure are outlined below:

Routing Computation Unit (RC) Failure A hard fault in the routing unit logic could cause all flits to be forwarded in the same direction or, in a more severe case, completely halt the generation of routing signals. The misdirection will not cause any data corruption, but it could lead to deadlock in deterministic routing algorithms. As soon as a failure in the RC unit is detected, it is broadcast to the adjacent routers. After knowing the failure status of the RC unit, the adjacent nodes can now substitute for the faulty RC unit by performing double routing, as shown in Fig. 4.5. Neighboring nodes sending flits to the faulty router need not worry, because their look-ahead routing will ensure that data arriving at the faulty node has already been taken care of. The problem affects the nodes receiving data from the faulty router; flits arriving at those nodes have not undergone look-ahead routing, because of the faulty RC unit in the previous router.

Therefore, nodes receiving flits from the faulty router must first conduct Current-Node Routing on those flits and then proceed to Look-Ahead Routing. The overhead involved is minimal and comes only from the few additional control signals; no additional resources are required.

Buffer Failure In a typical wormhole router, when a flit enters an input port, it is written to an input buffer queue. Bypassing the input buffer when the buffer is empty is a common optimization for performance; the flit heads straight to switch arbitration, and if it succeeds, the flit gets sent directly to the crossbar switch, circumventing the input buffers. This bypass path connecting the router input port with the crossbar input port can also be utilized in the event of buffer failure within a node. Virtual buffer management and switch allocation can still be performed in the current node, but buffer storage is offloaded to the previous node.

As soon as a flit stored in the previous node wins the switch arbitration in the current node, it can use the bypass path to circumvent the faulty buffer and proceed to the crossbar of the current node. In essence, data is physically stored in another router, but virtually queued and arbitrated in a different node through control signals between neighboring routers, as shown in Fig. 4.6. Under the Virtual Queuing mechanism, each buffer is not tied to a single VA arbiter in adjacent routers. Thus, a failure in a VA arbiter does not leave a particular buffer in a deadlock mode. There is a small latency penalty involving the round-trip delays of the handshaking signals, but it does avert the complete isolation of the faulty node. In terms of area cost, Virtual Queuing incurs minimal overhead, since no additional resource is required.

Virtual Channel Arbiter (VA) Failure Hard faults in this router-centric component are hardly recoverable by simply sharing of router resources. The operation of the VA cannot be offloaded to surrounding nodes, since its operation requires state information from several sources and it exhibits inter-dependencies with other router

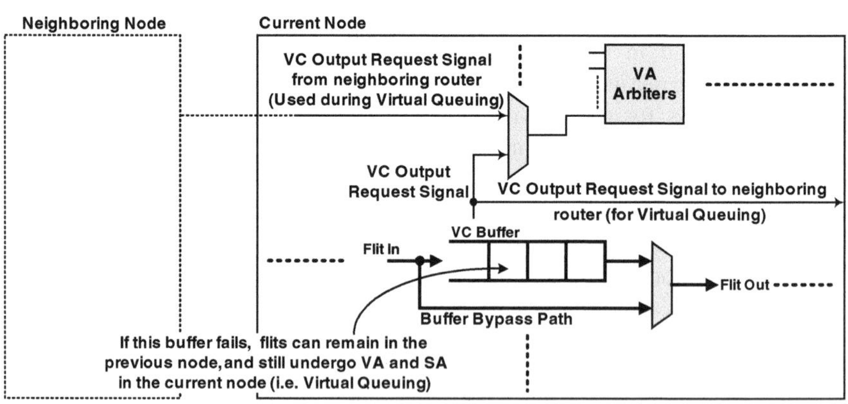

Fig. 4.6 The Virtual Queuing mechanism

elements and downstream nodes. Offloading such operations would require excessive transfer of state information. Faults in the VA can be bypassed only through resource replication, which is costly in terms of area and power overhead. The other option is to offload the arbitrations to the Switch Arbiter hardware, which contains identical arbiter modules. Nevertheless, this is infeasible because the SA is a per-flit component, meaning that it operates on all flits on a cycle-by-cycle basis. Since it is fully utilized, its operation cannot be preempted. The only choice, therefore, is to disable the whole router module in the event of a hard failure in the VA. However, whereas in generic architectures that would mean complete isolation of the entire node, in the proposed architecture only one of the two independent modules needs to be disabled.

Switch Allocator (SA) Failure Despite being a router-centric component, the SA can still be saved in the event of a hard failure in one of its components. Its operation cannot be transferred to neighboring routers because of the excessive transfer of state information required by such an endeavor. The solution proposed is much simpler and relies on the fact that the SA uses identical hardware with the VA, which is a per-packet component. Per-packet implies lower utilization, as explained in the beginning of this sub-section. This lends itself nicely to sharing of resources. By including a small number of compact 2-to-1 multiplexers at the input of some of the VA's arbiters, the SA can offload its operation to the VA, as shown in Fig. 4.7. Since the VA is operational only for header flit processing, its arbiters

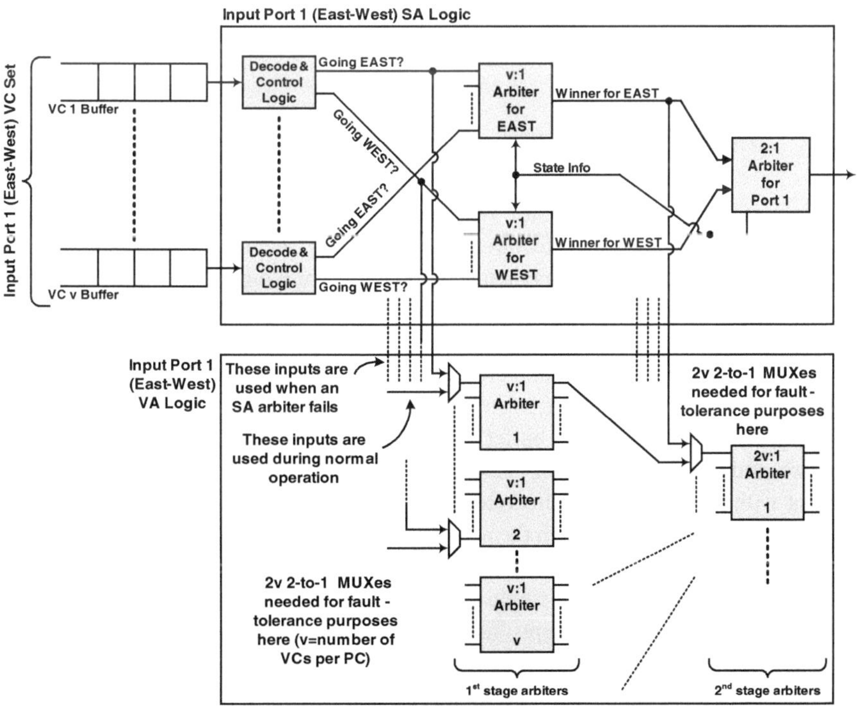

Fig. 4.7 Switch allocator fault-tolerance through resource sharing

can be used by the SA when they are idle. Performance, of course, is degraded because of the sharing of resources, but it is still a preferable alternative to the complete shutdown of the module (Row-Module or Column-Module). The area and power overhead imposed by the MUXes is minimal.

Crossbar and MUX/DEMUX Faults In the proposed RoCo router architecture, a decoder (DEMUX) is used to guide a flit into a group of path-sensitive queues, and a multiplexer (MUX) is used to direct a winning flit to the crossbar input. Therefore, the MUXes and DEMUXes all lie on the critical pathway of the router. A hard failure in one of these critical components can severely hamper the datapath progression. Once again, bypassing the datapath would imply replication of resources, which is not desirable. Hence, if any of these modules fails, the corresponding router module is blocked, while the other healthy module keeps operating.

4.5 Performance Evaluation

In this section, simulation-based performance evaluation of the RoCo architecture, a generic router architecture and the Path-Sensitive architecture of [73] is presented, in terms of network latency, energy consumption and fault-tolerance under various traffic patterns. The experimental methodology is described, and the procedure followed in the evaluation of these architectures is detailed.

4.5.1 Simulation Platform

A cycle-accurate NoC simulator was developed in order to conduct a detailed evaluation of the router architectures. The simulator operates at the granularity of individual architectural components, accurately emulating the major hardware components. The simulation test-bench models both the routers and the interconnection links, conforming to the implementation of various NoC architectures.

The simulator is fully parameterizable, allowing the user to specify parameters such as network size, topology, switching mechanism, routing algorithm, number of VCs per PC, number of PCs, buffer depth, PE injection rate, injection traffic-type, flit size, and number of flits per packet. The simulator models each individual component within the router architecture, allowing for detailed analysis of component utilizations and flit flow through the network. The activity factor of each component is used for analyzing power consumption within the network. We assume that link propagation happens within a single clock cycle. In addition to the network-specific parameters, our simulator accepts hardware parameters such as power consumption (dynamic and leakage) for each component and overall clock frequency. These parameters are extracted from hardware synthesis tools and back-annotated into the simulator for power profile analysis of the entire on-chip network.

4.5.2 Energy Model

The proposed RoCo router architecture, the Path-Sensitive router of [73] and a generic 2-stage 5-port router architecture were implemented in structural Register-Transfer Level (RTL) Verilog and then synthesized in Synopsys Design Compiler using a TSMC 90 nm standard cell library. The resulting designs both operate at a supply voltage of 1 V and a clock speed of 500 MHz. Both dynamic and leakage power estimates were extracted from the synthesized router implementation, assuming a 50% switching activity. These power numbers were then imported into our cycle-accurate network simulator for power analysis.

4.5.3 A Performance, Energy, and Fault-Tolerance (PEF) Metric

Traditional performance metrics used in NoC analysis, such as the Energy-Delay Product (EDP) and Power-Delay Product (PDP), focus on the two fundamental notions of latency (i.e., performance) and energy/power consumption. These metrics, however, do not capture the importance of reliability and its relation to both performance and power. Given that reliability is becoming a major concern in deep sub-micron technologies, it is imperative that evaluation of NoCs accounts for such issues. To address this need, we propose a composite metric which unifies all the three components: latency, energy, and fault-tolerance. Before introducing the new metric, we define three related terms.

Network Latency This is defined as the average number of cycles taken for end-to-end packet traversal, i.e., from a source to a destination.

Energy Consumption per Packet This is divided into two components: dynamic and leakage energy consumption. Both are defined as the total dynamic (or leakage) energy consumed in the network fabric over a time period divided by the total number of packets delivered during that period. Leakage power captures the effect of blocking delay, which translates into buffer static energy consumption. Dynamic power captures the effect of high contention within the router, which increases energy consumption due to excessive iterative operation of the SA and the VA units.

Packet Completion Probability This is defined as the number of received messages divided by the total number of injected messages into the on-chip network.

The inter-dependence between speed, power and fault-tolerance highlights the importance of a metric which can identify the best tradeoffs between these three competing traits. Hence, we introduce the Performance, Energy and Fault-tolerance (PEF) metric, as a comprehensive parameter that reflects the correlation between the three desired design goals. We define PEF as

$$PEF = \frac{(Average\ \ Latency) \times (Energy - per - Packet)}{Packet\ \ Completion\ \ Probability} = \frac{Energy - Delay - Product}{Packet\ \ Completion\ \ Probability} \qquad (4.2)$$

In a fault-free network, *Packet Completion Probability* = 1; thus, PEF becomes equal to EDP. Hence, PEF integrates reliability into EDP, thus providing a more complete evaluation metric.

4.5.4 Performance Results

The performance of the proposed RoCo router was analyzed and compared to two other existing router architectures (generic router, Path-Sensitive router of [73]) using the cycle-accurate simulator. All architectures were evaluated using an 8×8 2D mesh network. In the generic router architecture, three VCs per port were assumed, with a 4-flit deep buffer per VC; for a 5-port router, this configuration gives a total buffer capacity of 60 flits per router. To ensure fairness, since both the proposed RoCo architecture and the Path-Sensitive router have four ports instead of five, three VCs per port were assumed in both implementations, each with a 5-flit deep buffer; this gives a total buffer capacity of 60 flits per router, similar to the generic case. Each simulation consists of two phases: a warm-up phase of 20,000 packet injections, followed by the main phase which injects 1,000,000 additional packets. Each packet consists of four 128-bit flits. Under normal conditions, the simulation terminates when all packets are received at the destination nodes. In faulty environments, the simulation terminates after a long period of inactivity has elapsed (twice the time required to complete the simulation of a fault-free network).

Several experiments were conducted to evaluate the performance of all architectures under various traffic patterns and three different routing algorithms. The experiments employed uniform and transpose [82] traffic, and two synthesized workload traces, self-similar web traffic [85] and MPEG-2 video multimedia traces [86], in three different routing algorithms: DOR (XY routing), oblivious XY–YX routing, and minimal adaptive routing schemes. The results for multimedia traffic are not included here due to space constraints. Average network latency and power consumption were recorded for all experiments. Furthermore, several experiments were conducted to evaluate performance in faulty environments. A number of router faults (both Message-Centric and Router-Centric, as explained in Section 4.4) were randomly injected into the network infrastructure and the packet completion probability was analyzed. The traffic injection rate in these faulty networks was 30%. The latency, energy and fault-tolerance results were subsequently integrated into the PEF metric of Section 4.5.3 to reflect the combined measure.

The latency results of all three architectures for various traffic patterns are illustrated in Fig. 4.8 through 4.10. Clearly, the proposed RoCo router outperforms both the generic and Path-Sensitive routers in all traffic patterns and routing algorithms. With deterministic routing, the RoCo router reduces average latency by up to 35% compared to the generic router and by about 7% compared to the Path-Sensitive router. With XY–YX routing, these numbers become 38% and 10%, respectively. Finally, in adaptive routing, latency reduces by up to 40% compared to the generic router, and about 4% compared to the Path-Sensitive router.

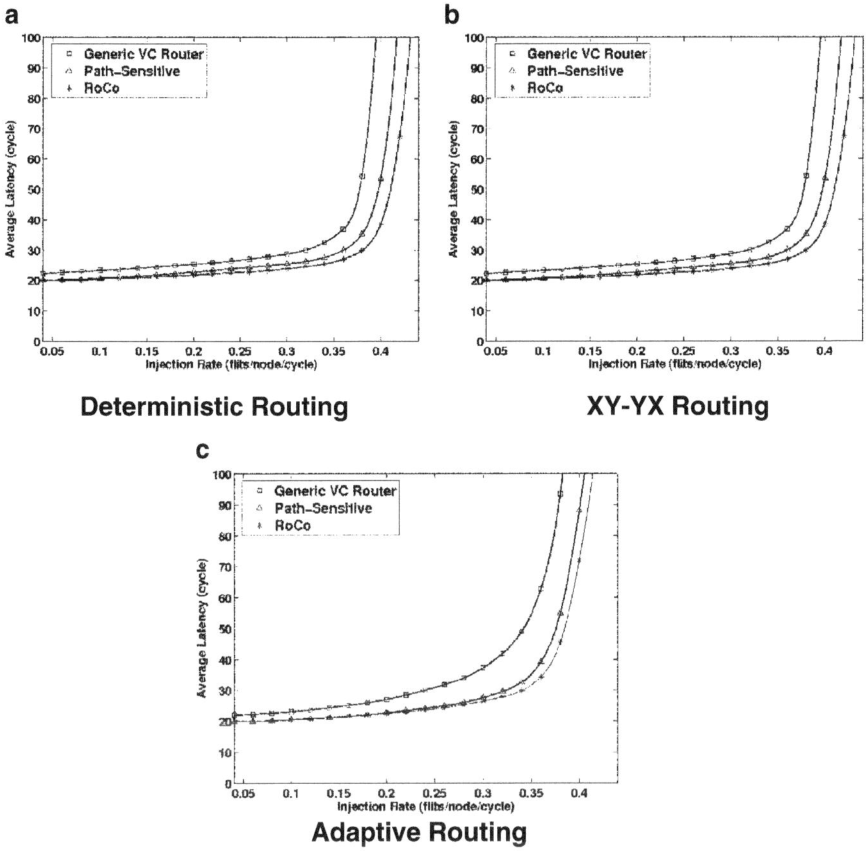

Fig. 4.8 Uniform random traffic: (**a**) deterministic routing, (**b**) XY–YX routing, (**c**) adaptive routing

The decoupling of the architecture into two distinct and functionally independent modules significantly reduces contention probability within the router. This effect manifests itself in lower average latency within the network. Furthermore, the use of the novel Mirroring Effect in switch arbitration increases crossbar utilization and reduces blocking.

Figure 4.13 compares the energy efficiency of the three different router architectures at 30% injection rate. The energy per packet is about 20% lower in the RoCo router, as compared to the generic router architecture, and about 6% lower compared to the Path-Sensitive router. This is a consequence of the simpler crossbars, smaller VA and SA units, and shorter logic depth. Therefore, the benefits afforded by the RoCo router are two-fold: reduced average network latency and lower energy consumed per packet. This is a testament to the fact that a more streamlined architecture can benefit both performance and power consumption.

Figures 4.11 and 4.12 illustrate the packet completion probabilities of the three router architectures when operating in faulty environments with 1, 2 and 4 random

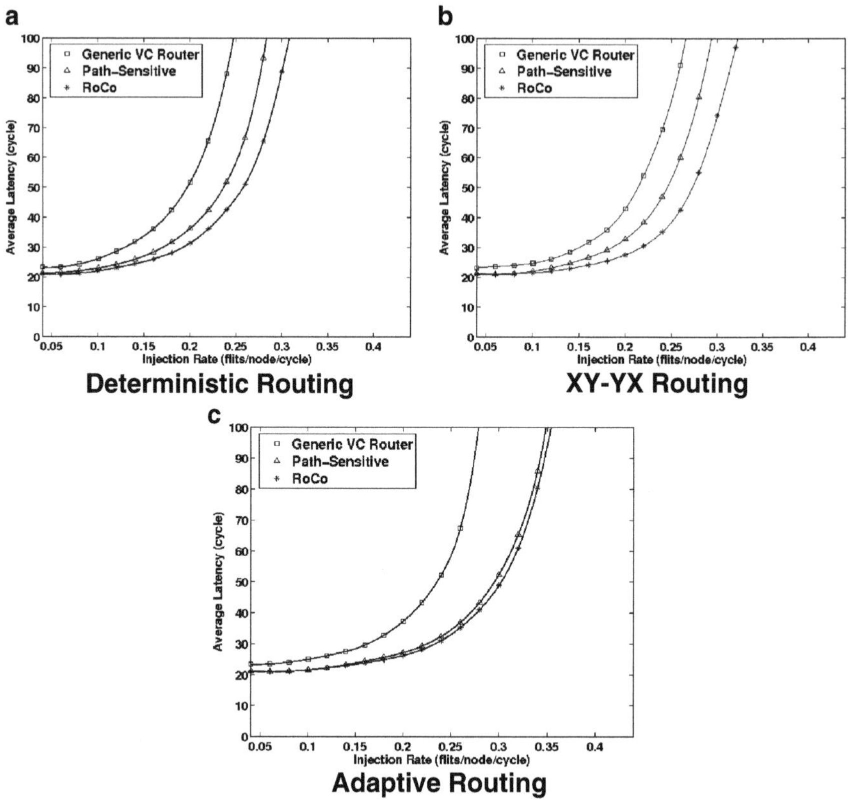

Fig. 4.9 Self-similar traffic: (**a**) deterministic routing, (**b**) XY–YX routing, (**c**) adaptive routing

network faults. Figure 4.11 concentrates on Router-Centric faults. These are critical faults, which cause the entire node to be blocked in the generic and Path-Sensitive cases. In the RoCo architecture, however, such faults only cause one of the two modules (Row-Module or Column-Module) to be blocked, thus, allowing for partial operation of the faulty router. Completion probability is consistently higher in the RoCo router in Fig. 4.11. As the number of faults increases from 1 to 4, the advantage of the proposed router becomes more obvious. The RoCo router provides up to 70% improvement in packet completion probability for different fault patterns with deterministic routing. The improvement drops to about 7% when adaptive routing is used. In both XY–YX and adaptive routing, the results are close because the routing algorithms provide alternate paths for all three architectures. However, this metric alone does not reflect the fact that even though completion probability is high in the generic and Path-Sensitive cases, the latency penalty incurred by excessive congestion around the faulty nodes is very high. This result will be captured later on in the PEF metric (Fig. 4.13).

Figure 4.12 focuses on Message-Centric faults, which are not critical. In the generic and Path-Sensitive routers, such faults would still cause the entire node to

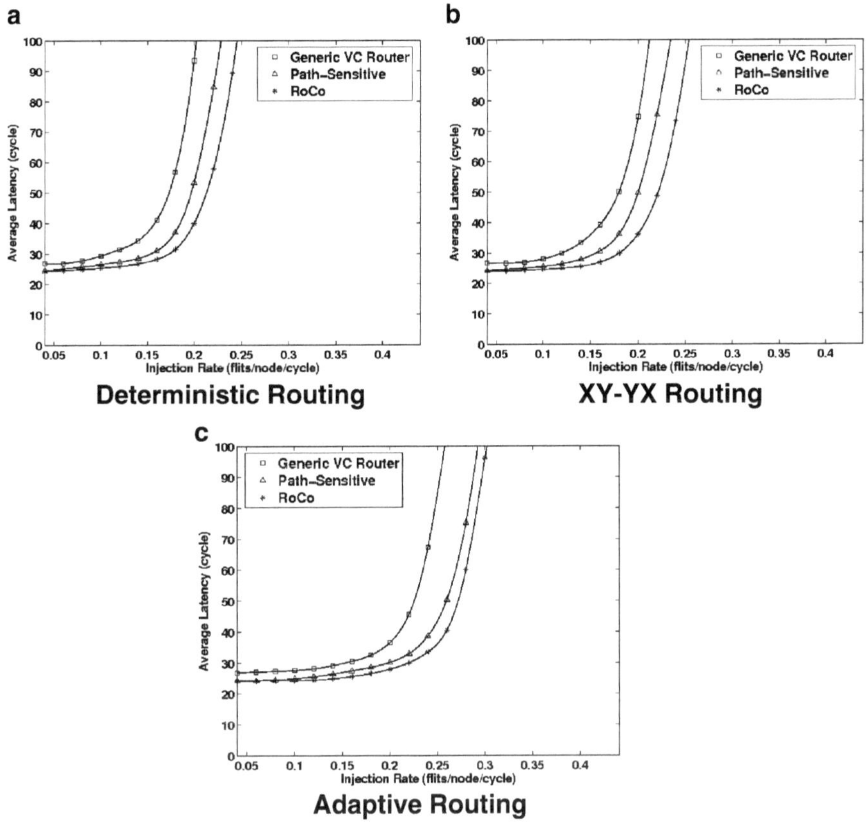

Fig. 4.10 Transpose traffic: (**a**) deterministic routing, (**b**) XY–YX routing, (**c**) adaptive routing

be blocked. However, in the RoCo router, such faults are remedied by the Recycling Mechanism of Section 4.4, which bypasses the faults through resource sharing. The oblivious routing schemes, i.e., deterministic and XY–YX, suffer more in the presence of faults, because of their rigid routing policies. These results indicate that the proposed Recycling Mechanism improves completion probability considerably without any significant router area overhead. Furthermore, even during critical Router-Centric faults, partial operation of the router can still serve network traffic in one dimension, thus alleviating congestion around the faulty node. This is achieved without any additional overhead. The results indicate that the RoCo router can achieve packet completion probabilities in oblivious routing that are close to those of adaptive routing schemes. This is of profound importance, since it indicates that the RoCo router provides uniform fault-tolerance under all routing algorithms. Through the recycling of faulty components and resource sharing, our proposed architecture degrades gracefully in faulty environments.

Figure 4.14 shows the combined measure (PEF) results for the three router architectures. The bars use the scale on the left-hand axis, while the curves use the scale

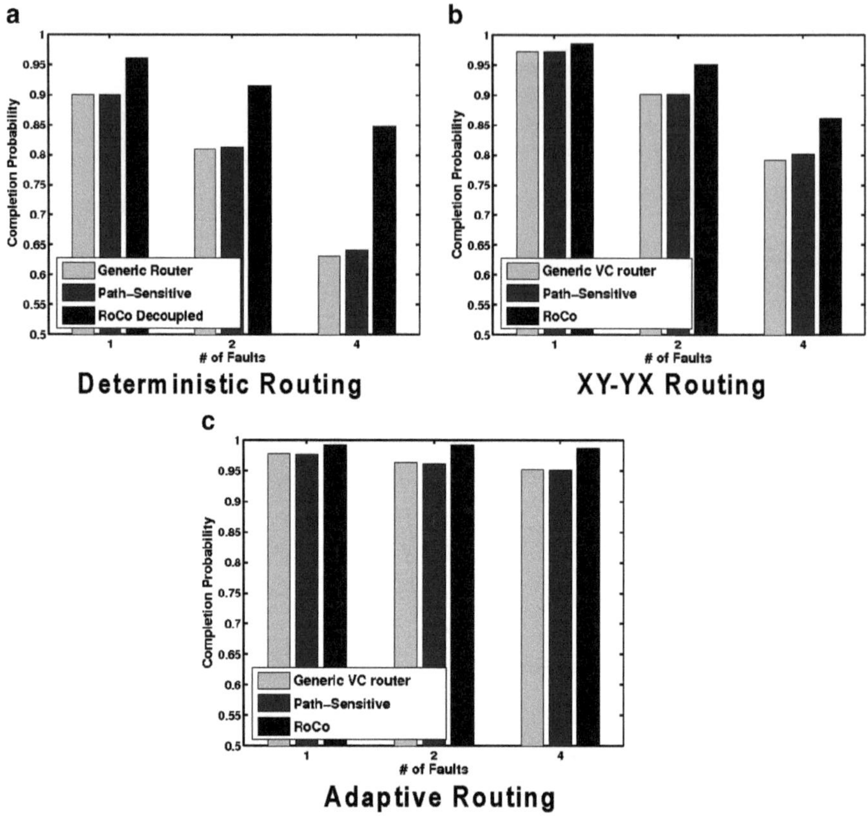

Fig. 4.11 Packet completion probabilities under various faults within router-centric and critical pathway components: (**a**) deterministic routing, (**b**) XY–YX routing, (**c**) adaptive routing

on the right-hand axis. This metric can successfully capture the subtle fact that despite high completion probabilities with adaptive routing, the generic and Path-Sensitive routers suffer from high latency due to congestion created around the faulty nodes. The RoCo router, on the other hand, has significantly lower latency numbers due to graceful degradation and the novel hardware recycling mechanism. Taking into consideration performance, energy consumption, and fault-tolerance in the integrated PEF metric, the RoCo router turns out to be the clear winner compared to the other two architectures. It provides almost 50% improvement compared to the generic router and 35% improvement compared to the Path-Sensitive router.

4.6 Chapter Summary

In this chapter, the author has presented a new router architecture, called Row–Column Decoupled Router, suitable for on-chip interconnects. The uniqueness of the proposed router is that it considers the three desirable objective functions:

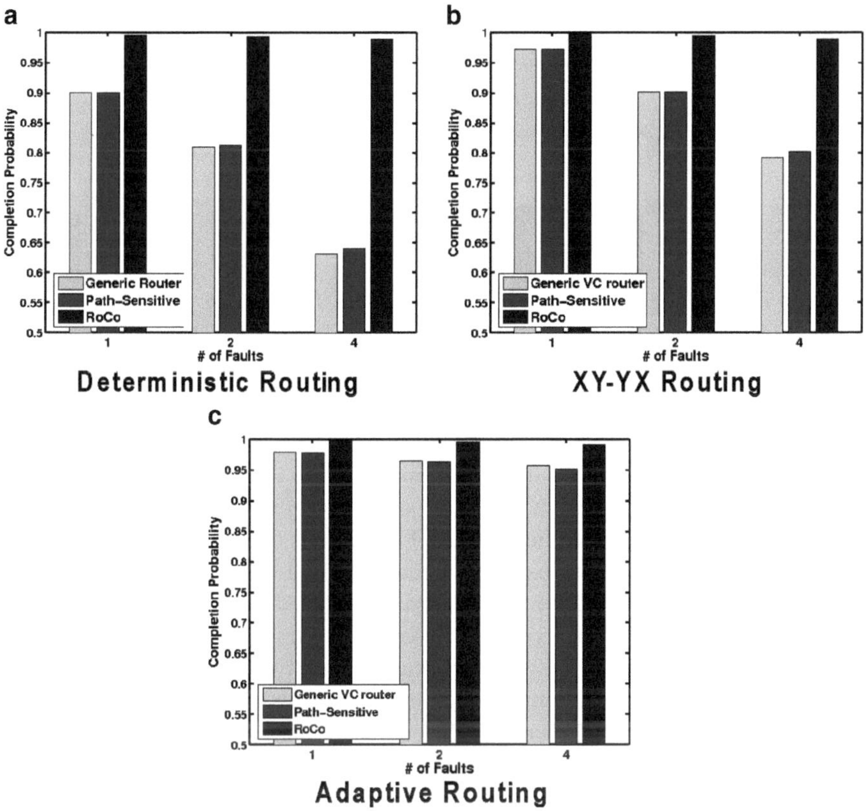

Fig. 4.12 Packet completion probabilities under various faults within message-centric and non-critical pathway components: (**a**) deterministic routing, (**b**) XY–YX routing, (**c**) adaptive routing

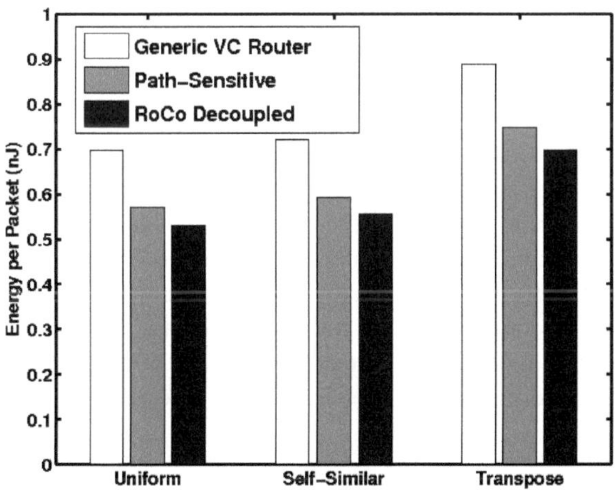

Fig. 4.13 Energy per packet

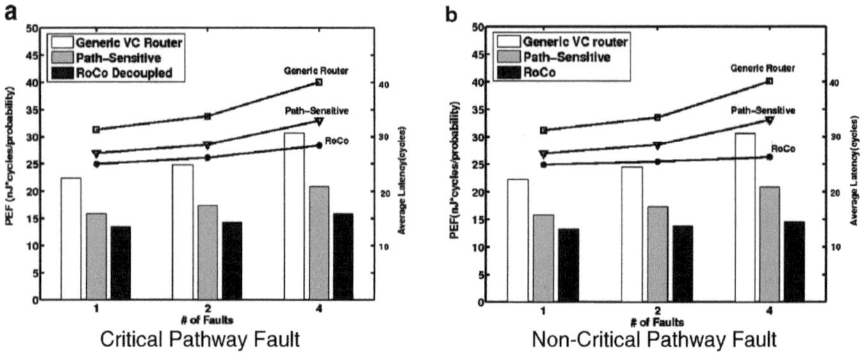

Fig. 4.14 Performance-energy-fault (PEF) product: (**a**) critical pathway fault, (**b**) non-critical pathway fault

performance, energy and fault-tolerance, in exploring the design space. The proposed 2-stage wormhole-switched RoCo router has a number of features that make it distinct compared to the earlier designs. First, it uses two smaller 2×2 crossbars instead of a larger 5×5 crossbar that is traditionally used for 2D mesh networks. Second, it uses a path-sensitive buffering scheme, where, the virtual channels are divided into four sets to support dedicated row and column routing in the two crossbars. These two features along with early ejection, mirrored allocation, look-ahead routing and speculative path selection help in reducing the contention. Third, unlike most earlier designs, it has been shown how deterministic (XY) routing, XY–YX routing and adaptive routing can be supported in this architecture. Fourth, because of the modular design, it has been shown how different types of faults such as VA, SA, and crossbar failures can be handled with graceful degradation, thereby providing better fault-tolerance compared to earlier designs. In addition, while all prior NoC studies have analyzed at best two of the three parameters, such as energy-delay product, this study introduces a comprehensive parameter, called PEF, for analyzing the performance, energy and fault-tolerance attributes of NoC architectures.

A flit-level, cycle-accurate simulator along with a detailed energy model for 90 nm synthesis were used to analyze the three objective functions using a variety of traffic patterns. Performance analysis with an 8×8 mesh network shows that the proposed router can reduce the average network latency up to 40% compared to a generic 2-stage router and by 10% compared to the Path-Sensitive router. In terms of energy consumption per packet, the proposed RoCo design outperformed the 2-stage router and Path-Sensitive router by 20% and 6%, respectively. The packet completion probability is improved by about 70% with deterministic routing. Evaluation with the composite performance, energy and fault-tolerance parameter (PEF) indicates that the RoCo architecture provides 50% and 35% better results compared to the generic and Path-Sensitive models, respectively.

Chapter 5
Exploring FaultoTolerant Network-on-Chip Architectures [37]

The advent of deep sub-micron technology has exacerbated reliability issues in on-chip interconnects. In particular, single event upsets, such as soft errors, and hard faults are rapidly becoming a force to be reckoned with. This spiraling trend highlights the importance of detailed analyses of these reliability hazards and the incorporation of comprehensive protection measures into all NoC designs. In this chapter, the author examines the impact of these transient and permanent failures on the reliability of on-chip interconnects and develops comprehensive countermeasures to either prevent or recover from them. In this regard, several novel schemes are proposed to remedy various kinds of soft and hard error symptoms, while keeping area and power overhead at a minimum. The proposed solutions are architected to fully exploit the available infrastructures in an NoC and enable versatile reuse of valuable resources. The effectiveness of the proposed techniques has been validated using a cycle-accurate simulator.

5.1 Introduction and Motivation

Aggressive technology scaling has accentuated the issue of reliability due to a rapid increase in the prominence of permanent faults; these are mostly caused from accelerated aging effects, such as electromigration, and manufacturing and testing challenges. Furthermore, soft upsets caused by cross-talk, coupling noise and transient faults are also a concern to overall reliability. The growing concern about reliability has prompted extensive research in this area. Many researchers [35,36,87–91] have proposed solutions for various individual aspects of on-chip reliability, such as soft faults and handling of hard failures within the network. Nevertheless, a comprehensive approach encompassing all issues pertaining to NoC reliability has yet to be made. In this chapter, the author proposes a comprehensive set of techniques to protect against the most common sources of failures in on-chip interconnects (including link errors, single-event upsets within the router, and hard failures). The proposed mechanisms incur minimal overhead while providing fool-proof

C. Nicopoulos et al., *Network-on-Chip Architectures: A Holistic Design Exploration*, Lecture Notes in Electrical Engineering 45,
DOI 10.1007/978-90-481-3031-3_5, © Springer Science+Business Media B.V. 2009

protection. Moreover, the new schemes cleverly employ resource sharing techniques to minimize the overhead imposed by the additional hardware.

To ensure protection from link errors due to crosstalk and capacitive loading, a flit-based hop-by-hop (HBH) retransmission scheme and the corresponding retransmission architecture are presented. With a minimal latency overhead of three clock cycles in the event of an error, this scheme successfully addresses the problems afflicting one of the most vulnerable components of an on-chip network, the inter-router link. In addition to providing link protection, the same architectural framework is also employed in a newly proposed deadlock-recovery scheme. While prior work in deadlock recovery has assumed additional dedicated resources, our technique uses existing retransmission buffers instead. This way, utilization of resources is maximized without incurring additional overhead.

While combinational logic circuits have traditionally been considered less prone to soft errors than memory elements, rapidly diminishing feature sizes and increasing clock frequencies are exacerbating their prominence. In fact, recent studies predict that the soft error rate (SER) per chip of logic circuits will become comparable to the SER per chip of unprotected memory elements by 2011 [92]. This observation has profound impact on the reliability of on-chip routers in the near future. The lack of protection from logic errors implies that soft errors afflicting a router's logic would escape the error detecting/correcting measures because they do not actually corrupt the data, but, instead, cause erroneous behavior in the functionality of the routing process. Therefore, it is imperative to provide robust protection against such upsets. A recent study [41] has addressed the issues of single-event upsets in the logic of individual hardware components. However, the proposed techniques were applicable to a specific type of router architecture. In this work, the author analyzes the intricacies of intra-router logic errors and provides comprehensive solutions relevant to all router architectures. The possible symptoms of logic errors in each module in the router pipeline are analyzed and detailed recovery mechanisms are provided for each case. The exploration of this topic culminates with the introduction of a novel Allocation Comparator (AC) unit, which provides full error protection to the virtual channel and switch allocation units at minimal cost.

Another well established topic in the design of reliable networks is fault-tolerant routing algorithms [35,36,87,88]. A fault-tolerant algorithm should be able to avoid faulty nodes and inter-router links to ensure correct delivery of messages. However, complex routing algorithms which require global knowledge of the network status are not viable in on-chip environments, where silicon real-estate and power are at a premium, and ultra-low latency is of paramount significance. In this chapter, a simple Proximity-Aware (PA) routing algorithm is proposed, which dynamically adapts its routing strategy to bypass faulty nodes. The algorithm employs a look-ahead snooping mechanism which collects health information from nodes up to two hops away. The look-ahead technique avoids faulty regions by rerouting messages to a safer area well in advance; this avoids unnecessary entanglement into highly faulty regions, thereby optimizing routing performance. Unlike other deadlock recovery algorithms [71], the proposed technique can also guarantee deadlock freedom in the presence of faulty nodes or links. This algorithm is then modified to detect and avoid possible temperature hot-spots in the network by observing flit activity in the router. Since high temperatures can adversely affect performance and reliability [25], thermal-aware routing becomes a useful protection mechanism. Simulation results show that our proposed proximity and thermal-aware routing algorithm (TPA) distributes network traffic more evenly than generic routing algorithms, mitigating hotspot phenomena.

Finally, a new routing paradigm called Service-Oriented Networking (SON) is introduced. Instead of routing packets based on destination address, the author proposes routing based on destination type. Recent trends in SoC design indicate that future systems will contain a number of identical processing elements on a single chip [3,4,93]. This characteristic can be exploited to provide fault tolerance against hard failures. Thus, even if a PE fails, SON can redirect messages to another PE which provides identical functionality. SON can dynamically reroute traffic in the event of a hard failure without service disruption. Simulation results validate the efficiency of this scheme even under multiple hard failures.

5.2 Simulation Platform Preliminaries

A cycle-accurate network simulator was developed to conduct detailed evaluation of the proposed schemes. The simulator operates at the granularity of individual architectural components, accurately emulating their functionalities. The simulation test-bench models the pipelined routers and their interconnection links. All simulations were performed in a 64-node (8×8) MESH network with 3-stage pipelined routers. Each router has five physical channels (PCs) including the PE-to-router channel, and each PC has a set of three associated virtual channels (VCs). One message (or packet) consists of four flits. The simulator keeps injecting messages into the network until 300,000 messages (including 100,000 warm-up messages) are ejected. A uniform message injection traffic pattern was used, where a node injects messages into the network at regular intervals specified by the injection rate. For a destination node selection, three distributions are used: normal random (NR), bit-complement (BC), and tornado (TN) [73]. Single link traversal is assumed to complete within one clock cycle, thus eliminating the need for pipelined links (which would incur further power and area penalties).

To evaluate fault-tolerance in the network, various soft and hard faults were randomly generated both within the routers and on the inter-router links. The simulator was also used to calculate the area and power overhead of the proposed architectures. A generic 5-port router architecture along with all proposed modifications were implemented in structural Register-Transfer Level (RTL) Verilog and then synthesized in Synopsys Design Compiler using a TSMC 90 nm standard cell library. The resulting design operates at a supply voltage of 1 V and a clock speed of 500 MHz. Both dynamic and leakage power estimates were extracted from the synthesized router implementation. These power numbers were then imported into the cycle-accurate network simulator and used to trace the power profile of the entire on-chip network. Average message latency and energy per packet were measured and used as the performance and energy parameters.

5.3 Handling Link Soft Faults

Primarily two types of soft faults could upset the on-chip network infrastructure: link errors occurring during flit traversal from router to router, and intra-router errors occurring within individual router components. The latter will be discussed in Section 5.4. This section focuses on link errors, which are mostly caused by channel disturbances such as cross-talk, coupling noise and transient faults [94]. Link errors have been studied extensively by researchers, since they have so far been considered the dominant source of errors in on-chip network fabrics. They have been tackled within the context of two central themes – correction and retransmission. Some degree of error correction can be achieved through the use of Error Correcting Codes (ECC), as in [95,96]. These codes achieve what is known as Forward Error Correction (FEC). Similarly, retransmission schemes can also be used to compensate for link errors.

Hybrid techniques [90], which provide both error correction and retransmission, allow for more robust protection of data. Hybrid solutions compensate for the limitations of error correcting codes. For example, Single Error Correction and Double Error Detection (SEC/DED) codes can correct at most one error, but can detect double-bit errors. Therefore, upon detection of a double-bit error, the SEC/DED unit may invoke a retransmission mechanism. Retransmission can occur in two different forms: End-to-End (E2E) or Hop-by-Hop (HBH). In an E2E scheme, the original data is checked only at the destination node, while in an HBH scheme, data is checked in all routers along the path from a source to a destination. Both flavors require dedicated buffers, as opposed to FEC techniques, but they can handle multiple-bit errors, since a clean copy of the data is always maintained.

Both FEC techniques and E2E retransmission schemes suffer severely from errors in the header flit. For example, if the destination node address of a packet is corrupted during the transfer, the packet might be routed to a wrong destination. Even if FEC can correct the error at the (wrong) destination node, the packet should be sent to the correct destination creating additional network traffic. Similarly, E2E schemes need to send a retransmission request from the wrong destination to the source node. Moreover, if the source node address is corrupted, E2E techniques cannot send the retransmission request to the correct source. Thus, it is very important to keep the header information from being contaminated even if the probability of header error is small, reference [41] addressed this problem by adopting HBH header error checking in both FEC and E2E schemes. Figure 5.3 shows that E2E schemes suffer from prohibitive latency penalties as error rates increase. E2E schemes also require larger retransmission buffers to account for worst case round-trip delay between a source and destination [41].

Considering all these aspects, HBH retransmission together with FEC seems to be the best choice to handle link faults. To that extent, a minimal-overhead flit-based

HBH retransmission scheme is proposed. The impact of the additional buffer overhead is mitigated by utilizing these same buffers for a newly-proposed deadlock recovery mechanism, discussed in Section 5.3.2.

5.3.1 Flit-Based HBH Retransmission Scheme

Our proposed minimal overhead HBH retransmission scheme requires a 3-flit-deep retransmission buffer per virtual channel, since a flit should be kept for 3 cycles after it leaves the current node. This 3-cycle delay corresponds to the sum of the link traversal delay (1 cycle), error checking delay at the adjacent receiving node (1 cycle), and the Negative Acknowledgement (NACK) propagation delay (1 cycle). The retransmission buffer is implemented as a barrel-shift register. This way, a flit is stored at the back of the buffer upon transmission on the link, and it moves to the front by the time a possible NACK signal arrives from the receiving node. The simplest type of transmission buffer is a First-In-First-Out (FIFO) buffer. Such an implementation has one input port and one output port, and involves simple control logic. The proposed architecture is shown in Fig. 5.1.

Conceptually, our proposed scheme works similar to the simple retransmission schemes described in [82,97,98]. However, in terms of implementation, references [82,97,98] use a single transmission buffer that contains both sent and unsent flits together, and use pointers to track their positions. This requires that every buffer slot has an exit port so that flits can be transmitted from the middle of the buffer. This complicates the logic and incurs wiring overhead. Further, they use both acknowledgement (ACK), as well as NACK signals, whereas our proposed scheme

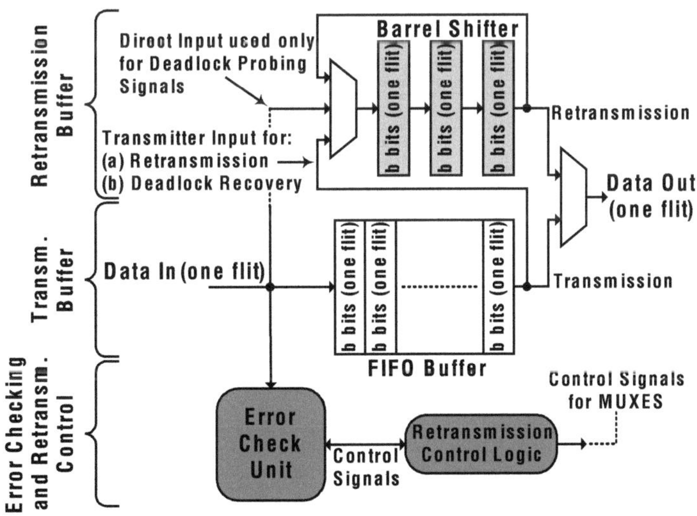

Fig. 5.1 Proposed transmission/retransmission buffer architecture

only sends NACK signals when an error is detected. Reference [98] uses link-level retransmission together with the Unique-Token Protocol (UTP) to ensure reliability. However, it requires at least two copies of a packet at all times in the network, increasing buffer occupancy and flow control complexity.

In case of a flit error in the proposed scheme, two subsequently arriving flits must be dropped until the correct flit arrives from the previous node upon retransmission. Once the correct flit is received, all previously dropped flits must then be retransmitted. This scenario is illustrated with a flit-flow example in Fig. 5.2. The example traces the operation of the HBH retransmission mechanism when the header flit H1 is corrupted during link traversal. A clean copy of H1 is stored in the retransmission buffer when H1 is sent to the link. The error check logic detects errors in H1 in the receiving node and sends a NACK signal to the transmitting node in the next clock cycle. As seen in Fig. 5.2, the receiving router drops the subsequent two flits (D2 and D3).

While this may seem an inefficient recovery method, it should be noted that a retransmission event will be highly unlikely under normal operation, since the architecture already employs a single-error correction scheme. The probability of a double (or higher) error within a single flit is low in on-chip networks. Furthermore, the corrected flit (H1 in the example of Fig. 5.2) arrives within 3 clock cycles, as explained at the beginning of this sub-section. This implies that only two flits need to be dropped during a retransmission event. Retransmission of these two flits incurs a latency penalty of only two clock cycles. Therefore, a possible latency improvement of two clock cycles does not warrant the implementation of a more complex architecture which would be able to handle in-situ re-arrangement of flits within each router. While such implementations are very common in macro

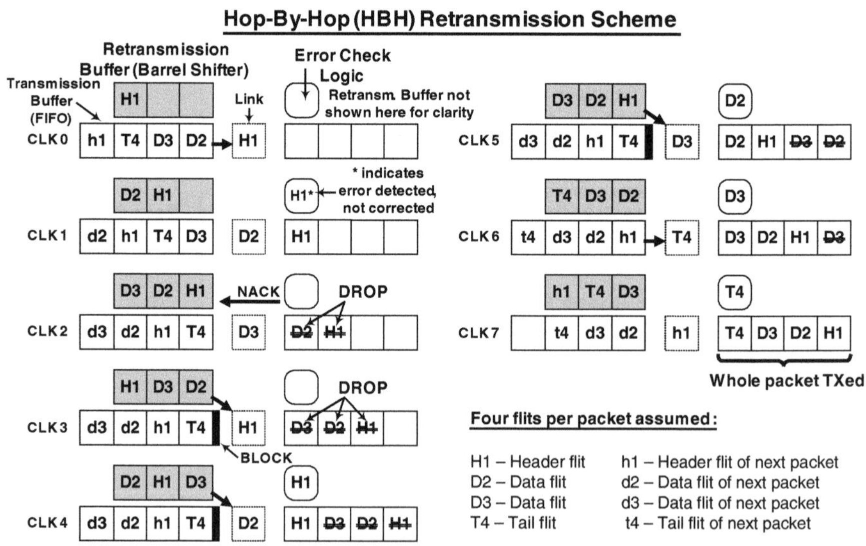

Fig. 5.2 Hop-by-hop (HBH) retransmission mechanism

networks, they are prohibitive in on-chip environments, because the latter have much stricter area and power budgets. The excessive area and power penalty imposed by these modifications, compounded by the increased wiring complexity, clearly overshadow the small improvement in latency during a low-probability retransmission event. Cycle-accurate simulation of the proposed scheme in an 8×8 network validates these assertions, as shown in Fig. 5.4. The retransmission scheme is so efficient that average latency remains almost constant even at 10% error rates. This behavior is a direct consequence of the minimal latency incurred during a retransmission, as shown in Fig. 5.2. Furthermore, retransmission occurs only between two adjacent hops; this restricts movement of retransmitted flits to a single inter-router link, which, in turn, has minimal impact on overall network traffic. Similarly, Fig. 5.5 illustrates the negligible effect of the proposed scheme on the

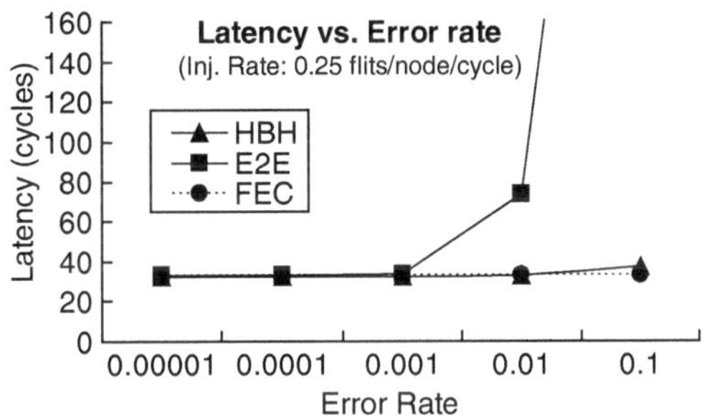

Fig. 5.3 Latency of various error-handling techniques

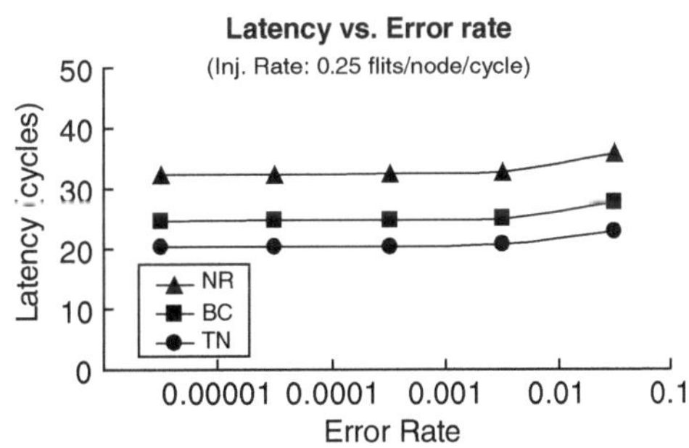

Fig. 5.4 Latency overhead of the HBH retransmission scheme

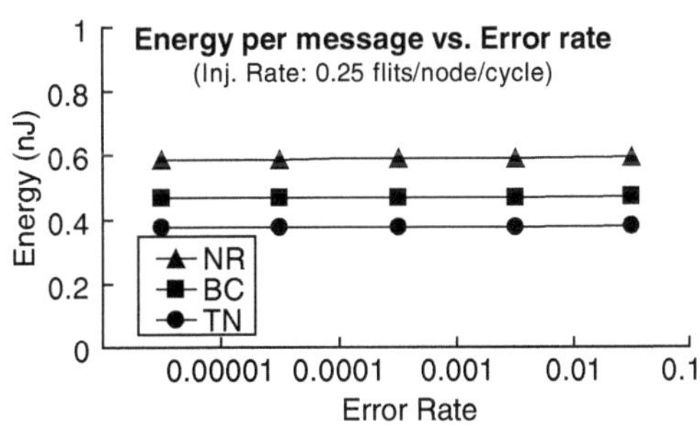

Fig. 5.5 Energy overhead of the HBH retransmission scheme

energy-per-packet metric. Since retransmission is done on a hop-by-hop basis for individual flits, the power overhead of a single-hop flit transmission is insignificant compared to the total power budget for complete packet traversal from source to destination.

It should be noted that the retransmission buffer also constitutes an essential component of our proposed deadlock recovery scheme, which is analyzed in detail in Section 5.3.2. Utilizing the same hardware for both schemes further subsidizes the area and power overhead incurred by the additional circuitry.

5.3.2 *Deadlock Recovery*

The concept of deadlock has been extensively researched in the literature. Some researchers use preventive schemes [99,100], while others propose recovery schemes [71,101]. However, these methods typically place constraints on resource use, preventing the system to work at full throttle. For example, an adaptive routing algorithm can use escape Virtual Channels (VCs) to recover from deadlocks, as described in [71]. The flits in these escape VCs, however, are managed by a deadlock-free deterministic routing algorithm, thereby limiting adaptivity.

Moreover, many of these techniques cannot guarantee deadlock freedom in a network with hard faults (router or link faults). They all assume a fault-free environment. This assumption, however, no longer seems reasonable in NoC environments where the probability of failure is relatively high. Several techniques have been proposed to address this issue in macro networks [82,83], but most of them adopt complex algorithms which are not suitable in resource-constrained environments like on-chip interconnects. Thus, it is imperative to provide simple, yet effective solutions to minimize performance degradation.

5.3.2.1 Proposed Deadlock Recovery Scheme

To address these issues, we propose a scheme, which (1), instead of using additional dedicated resources, utilizes the existing retransmission buffers to break deadlocks, and (2) provides deadlock recovery in both fault-free and faulty environments using a very simple retransmission-buffer management policy. Hence, through efficient resource sharing, we can transform the retransmission buffers into a multifaceted reliability component in our system.

The retransmission buffers can serve a dual purpose, mainly because they are used only when packets are being transferred from one node to another. As network traffic increases, packet blocking increases and, as a result, the utilization of the HBH retransmission buffers will decrease due to decreased flit transmissions. Figures 5.6 and 5.7 show the utilization of both transmission and retransmission buffers, respectively, for the adaptive (AD) and deterministic (DT) routing algorithms. In most cases, the utilization of the retransmission buffer does not follow that of the transmission buffer; instead, retransmission buffers are mostly

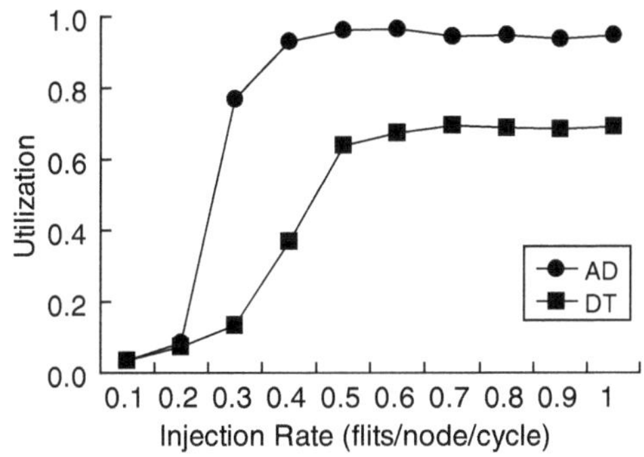

Fig. 5.6 Transmission buffer utilization

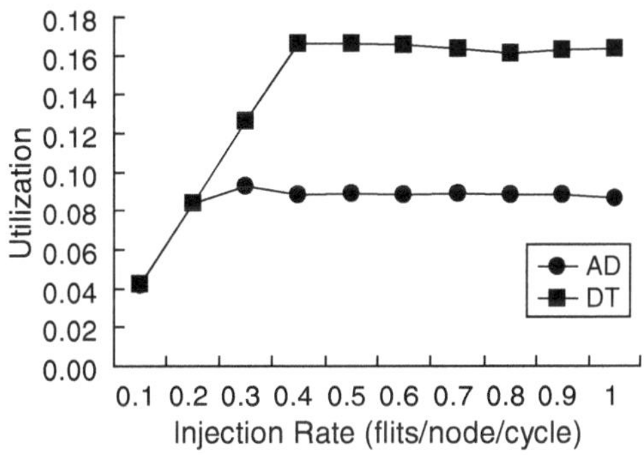

Fig. 5.7 Retransmission buffer utilization

underutilized. Furthermore, if a packet is permanently blocked due to a deadlock, the associated retransmission buffer will be empty, since there has been no data transmission for an extended period of time. Based on these observations, we propose a smart HBH-retransmission-buffer management scheme that exploits these idle buffers for deadlock recovery.

When a deadlock occurs, if any of the packets involved in the deadlock configuration can proceed by one buffer slot, all the other packets also involved in the deadlock can proceed as well. This can be achieved with the presence of a single empty buffer slot; if all packets continue to proceed in this fashion, the deadlock will eventually be broken, since some packets will ultimately move out of the deadlock configuration. In other words, instead of providing a dedicated escape channel to the destination node, as proposed in [71,101], our scheme gradually shifts flits without breaking the cyclic dependency, until the deadlock is broken.

For example, assuming that four nodes are involved in a deadlock configuration, we have (4×3) retransmission buffers (12 in total) that are empty. Therefore, if each node temporarily moves three flits from the normal transmission buffers to the retransmission buffers, it will create an additional available buffer space for the preceding node in the deadlock configuration. As soon as the buffer space becomes available, the flits in the retransmission buffer can be sent to the next router. Thus, flits will be able to advance, and after several iterations, some flits will move out of the deadlock configuration, thereby breaking the deadlock situation.

Figure 5.8 shows an example of this scenario in detail, where a packet consists of 4 flits and the normal transmission buffer can store up to 4 flits. In step 1, a deadlock is detected and flits are moved to the retransmission buffer, as shown in step 2. The additional buffer space created by this move allows flits in the retransmission buffers to be transmitted to the next nodes. Since the retransmission buffers in our proposed architecture are barrel shifters, transmitted flits also move to the back of the retransmission buffer (flits enclosed by a thick square), as shown in steps 3 to 5. Three clock cycles later, the retransmission buffer will be empty again, as shown in step 7. At this point, the buffer state is exactly the same as in step 1, except that every flit has advanced by 3 buffer slots. This procedure will be repeated until at least one of the packets breaks the deadlock by going out to a direction away from the deadlock configuration. Once the deadlock configuration is broken, each node resumes its normal operation. In the example of Fig. 5.8, we assume that all three nodes involved in the deadlock initiate deadlock recovery action simultaneously for the sake of clarity. However, deadlock recovery need not be synchronized, as long as all nodes eventually start deadlock recovery. The proposed probing technique described below will handle this asynchronous behavior.

The proposed scheme places a lower limit on buffer size to ensure correct functionality. The technique must account for the worst-case scenario, where partially transferred messages prevent other messages from entering the transmission buffers, and thus, absorption of these partially transferred messages is necessary during the deadlock recovery process, as illustrated in Fig. 5.9. Note that no new packets

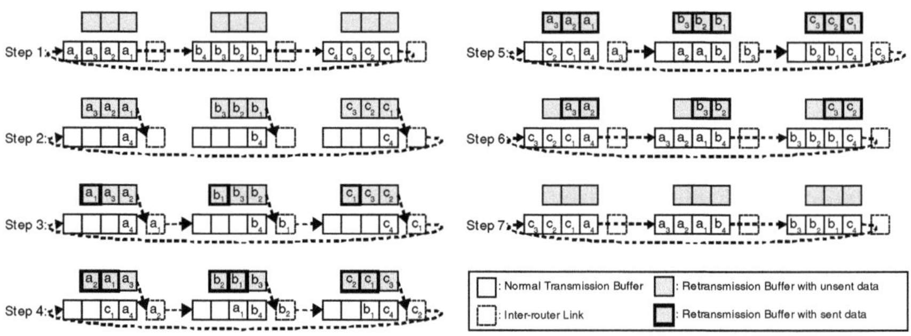

Fig. 5.8 Deadlock recovery example

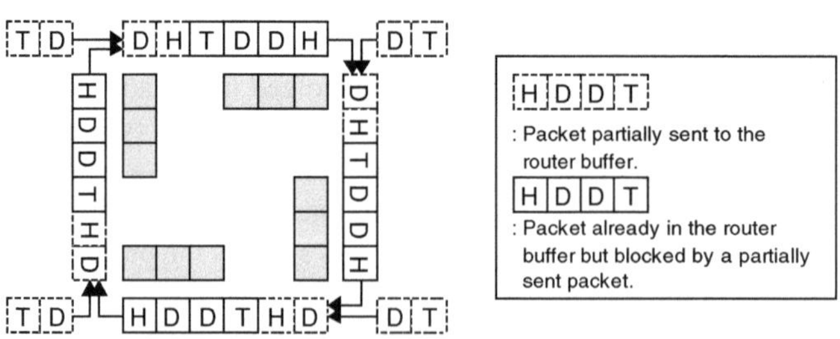

Fig. 5.9 Worst-case deadlock recovery example

are allowed to enter the transmission buffers that are involved in the deadlock recovery.

To handle the worst-case scenario, the total buffer size (i.e., transmission and retransmission buffers) must be large enough to accommodate the remaining flits of a partial packet and still have at least one empty slot.

Theorem: The proposed scheme ensures deadlock freedom if the buffer size is larger than the lower limit specified in Eq. (5.1).

$$B > M \times \sum_{i=1}^{n} N_i, \tag{5.1}$$

$$B = \begin{cases} B2 = T = \sum_{i=1}^{n} Ti, \textit{normal}\ \text{mode} \\ B2 = T + R = \sum_{i=1}^{n} (Ti + Ri), \textit{deadlockre}\ \text{covermode} \end{cases}$$

- B : Total buffer size (either B_1 or B_2)
- T, (T_i) : Total size of the Transmission buffer (at node i)
- R, (R_i) : Total size of the Retransmission buffer (at node i)
- n : Number of nodes involved in the deadlock
- M : Number of flits per packet (message)
- N_i : Maximum number of different packets in a transmission buffer at node i ($= \lceil T_i / M \rceil$)

Proof: When a deadlock is detected, the transmission buffers cannot accommodate any more flits, and, therefore, $B = B_1 \leq M \times \sum_{i=1}^{n} N_i$. At this point, the nodes switch to the deadlock recovery mode and if $B = B_2 > M \times \sum_{i=1}^{n} N_i$, then all the messages involved in the deadlock can be absorbed into the buffers (transmission + retransmission) with at least one empty slot still available. Since only packets involved in the deadlock can use this empty buffer slot(s), they can now proceed, and eventually, the network can recover from the deadlock.

Examples of the lower limit condition in deadlock recovery mode are shown below for the configurations of Figs. 5.8 and 5.9. Both examples show that they meet the minimum buffer requirement, and therefore, the deadlock can be broken.

$$Ti = 4, Ri = 3, M = 4, Ni = \lceil 4/4 \rceil = 1, n = 3$$
$$\therefore B = B2 = n \times (4+3) = 21 > 4 \times (n \times 1) = 12$$

$$Ti = 6, Ri = 3, M = 4, Ni = \lceil 6/4 \rceil = 2, n = 4$$
$$\therefore B = B2 = n \times (6+3) = 36 > 4 \times (n \times 2) = 32$$

If the retransmission buffers are not to be used for deadlock recovery, then this lower limit for the total buffer size is no longer necessary. Therefore, if we forego deadlock recovery support, only three retransmission buffers will be needed per VC for link error correction (see Section 5.3.1), regardless of the regular transmission buffer size.

5.3.2.2 Probing for Deadlock Detection and Recovery

To detect possible deadlocks, most of the previous approaches adopted a threshold value of blocked cycles, after which the router assumed that the blocked flit was involved in a deadlock. This approach is guaranteed to detect all possible deadlocks [71]. However, it can also give false positives, where a node assumes a deadlock even though the flit is simply experiencing long blocking delay. Increasing the triggering threshold value will decrease the number of false positives, but increasing the threshold value arbitrarily will cause the number of blocked flits in the network to increase. In order to predict the most appropriate threshold value, one needs to consider a multitude of parameters, such as the network traffic among nodes, the traffic load, the routing algorithm and the deadlock recovery scheme. This can be a daunting feat, since the exploration space is huge.

To overcome this limitation, we aim to formulate a different methodology, which will detect only actual deadlocks without any false positives; this optimizes network performance, while eliminating the need to precisely identify an optimal threshold value. We propose a probing technique, whereby a compact probing signal is sent along the suspected deadlock path after a flit has experienced more than a predefined number of cycles (C_{thres}) of blocking. The probe will check whether the flit is involved in a real deadlock or not. While the selection of C_{thres} will also affect network performance (as the threshold value described above), its impact is less pronounced because the probing technique will ensure that no action is prematurely taken. In other words, the threshold itself does not initiate deadlock recovery. The probing technique will first assess the situation to prevent the occurrence of any false positives. Therefore, the value of C_{thres} need not be precisely calculated; its effect on overall network performance will be minimal as long as the value chosen is not excessively high.

The proposed probing technique detects a deadlock based on the following two rules:

Rule 1: *After a flit experiences more than* $\boldsymbol{C_{thres}}$ *cycles of blocking, the router sends a probing signal to the next node specifying the VC buffer of the suspected flit.*
Rule 2: *When a node receives a probing signal, it checks the status of the buffer specified in the probing signal. If the VC buffer is also blocked in the current node or the node is in deadlock recovery mode, it forwards the probing signal to the next node, modifying the VC identifier accordingly. Otherwise, it discards the probing signal.*

If the probing signal returns to the original sender node, then the latter can safely assume that the flit under investigation is involved in a deadlock configuration; this is because the probing signal can return to the sender only if there is a cyclic path dependency and all intermediate nodes also experience blocking (by Rule 2). If increased blocking delay due to a hard failure causes a node to suspect deadlock, the subsequent probing signal will be discarded by the router adjacent to the faulty node, which will redirect blocked flits to another direction using an adaptive routing scheme, breaking the deadlock, if any.

After the probe returns, the sender sends an activation signal that triggers the nodes involved in the deadlock to switch to the deadlock recovery mode. The sender node switches to the deadlock recovery mode after the activation signal returns. To handle the case where multiple nodes in the same deadlock configuration send probing signals at the same time, we need two more rules:

Rule 3: *A node will discard an activation signal unless it has received a probing signal from the same sender node before.*
Rule 4: *If a node receives a valid activation signal (as per Rule 3), while it is waiting for its own probe to return, it switches to the deadlock recovery mode and discards its own probe when it finally returns, since the deadlock recovery mode has already been activated by another node involved in the same deadlock configuration.*

To avoid incurring any additional overhead in supporting dedicated probing lines, we propose using a regular flit transmission for the probing signal, which can use the retransmission buffers in each suspected node to propagate. Note that the retransmission buffers are empty in nodes experiencing long blocking. This will ensure that the probing signal itself will not be blocked in an intermediate router. Figure 5.1 shows how an incoming link can feed the retransmission buffer directly. Since the probing signal is a regular flit, it will also be protected by the error correcting blanket, thus ensuring its safe traversal through the network.

5.4 Handling Soft Errors in Intra-Router Logic

Until recently, soft errors were tackled within the context of memory cells or registers. This has led to the widespread use of error detection and correction circuits to protect memory arrays. Combinational logic circuits, on the other hand, have been found to be less susceptible to soft errors in equivalent device technologies due to the naturally occurring logical, electrical and latching-window masking effects [102]. However, decreasing feature sizes and higher operating frequencies are rapidly thinning the protective effect of these masking phenomena. As mentioned before, research has indicated an exponential increase in the soft error rate (SER) per chip of logic circuits in the future [92]. Hence, it is crucial that modern router designs account for these events to ensure reliable and uninterrupted operation of the on-chip network. The notion of logic errors resulting from soft error upsets is directly related to the number of pipeline stages within the router. While the proposed measures are the same for all implementations, the recovery process differs depending on the number of pipeline stages present (and, thus, the amount of speculation employed by the architecture). The following sub-sections discuss the effects of soft errors on each router component along with proposed counter-measures. The recovery process for the different pipeline implementations is also analyzed. The latency overhead in the cases of 2-stage and 1-stage routers assumes successful speculative allocation in the recovery phase. Mis-speculation will increase the overhead, but mis-speculation occurs during normal operation as well and is unpredictable.

5.4.1 Virtual Channel Arbiter Errors

The VA, like the routing unit, operates only on header flits. All new packets request access to any one of the valid output VCs, returned by the routing function. The VA arbitrates between all those packets requesting the same output VC. The VA maintains states of all successful allocations through a pairing between input VCs and allocated output VCs. It is this state that effectively opens up the "wormhole" for all subsequent flits of the same packet. Soft errors within the VA may give rise to four different scenarios:

(1) One input VC is assigned an invalid output VC: For example, suppose a PC has 3 VCs – designated by 00, 01, and 10. A soft error might cause the assignment of invalid VC 11. Such an assignment will block further traversal of the packet through the network.
(2) An unreserved output VC is assigned to two different input VCs: This will lead to packet mixing, and, eventually packet/flit loss. Flits from both packets will follow the same wormhole, since they are seen as one packet by the routers. As soon as the tail flit of one of the two packets releases the wormhole, any subsequent flits of the other packet will essentially be stranded in the network. For example, incoming packets from the North and West both can be assigned the same output VC in the South.
(3) A reserved output VC is assigned to a requesting input VC: This case is very similar to case (2) above. The new packet will erroneously follow the existing wormhole, following a path to a wrong destination. The same consequences will result as above.
(4) An erroneous, yet unreserved, output VC is assigned to a requesting input VC: In this scenario, there are two different types of erroneous output VC assignments:
    (a) The wrong output VC belongs to the intended PC. This is a benign case, since the packet will still be forwarded to the same physical direction as originally intended.
    (b) The wrong output VC belongs to a PC other than the intended one. This case is similar to the misdirection situation to be analyzed in Section 5.4.2. It may lead to deadlock in deterministic routing algorithms. The solution described in Section 5.4.2 will also protect against this type of error.

The proposed safeguard for VA logic errors is the addition of a compact hardware unit to the router architecture, called the Allocation Comparator (AC). The proposed AC unit is shown in Fig. 5.10. The unit employs purely combinational logic, in the form of XOR gates, to compare the RT state entries and the VA state entries. The AC unit performs three types of comparisons in parallel, within one clock

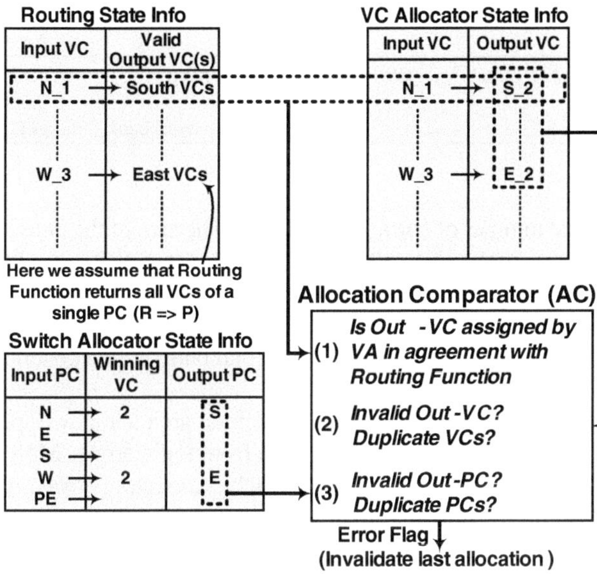

Fig. 5.10 The proposed Allocation Comparator (AC) unit

cycle. It first checks to see if the output VCs assigned by the VA unit are in accordance with the output of the routing function (i.e., RT unit). For instance, if a soft error causes the VA to erroneously assign an output VC in the North PC, while the RT unit had indicated the assignment of a VC in the South PC, the AC unit will trigger an error flag, thus invalidating the VA allocation of the previous clock cycle. This comparison protects against scenario (4b) above. Secondly, the AC unit checks the VA state info to detect both invalid and duplicate output VC assignments. Should any of these cases appear, an error flag is raised. This comparison safeguards against scenarios (1) through (3) above. Finally, the AC unit checks for Switch Allocation errors, but this is discussed in the following sub-section.

The duration of the recovery phase is independent of the pipeline architecture. In all cases except the 4-stage router, parallelization implies that the AC unit will operate in the same stage as the crossbar traversal (i.e., right after the VA operation). This means that if an error is detected by the AC unit, a NACK should be sent to all neighboring routers to ignore the previous transmission. Then the previous VA allocations are repeated in the current router, thus incurring single-clock latency overhead. In a 4-stage router, the AC unit will detect the error by the end of stage 3 (i.e., before crossbar traversal); therefore, no erroneous transmission will occur. The latency delay is still one clock cycle.

While adding additional hardware increases the overall area and power consumption of the router, the proposed unit was deliberately architected to be as small and efficient as possible. First, the number of state entries to be compared is equal to PV, where P is the number of input/output ports and V is the number of VCs per port. For a typical 5-port mesh NoC router (North, East, South, West, and PE) with

Table 5.1 Power/area overhead of allocation comparator (AC) unit

Component	Power	Area
Generic NoC Router (5 PCs, 4 VCs per PC)	119.55 mW	0.374862 mm^2
Allocation Comparator (AC)	2.02 mW (+1.69% overhead)	0.004474 mm^2 (+1.19% overhead)

4 VCs per PC, the number of entries is $5 \times 4 = 20$. The size of the entries is minimal, since the VC IDs are only a few bits long (e.g., 2 bits for 4 VCs per PC). Thus, the data to be compared is very small. To validate the architecture's compactness and efficiency, we synthesized the comparator unit in 90 nm technology. The area and power budgets of the unit, as compared to the total budget for a generic NoC router, are shown in Table 5.1.

It is evident that the AC unit imposes a minimal area and power penalty on the overall design, while providing full protection from logic errors. Moreover, the AC unit is also used to protect against Switch Allocator errors, described in Section 5.4.3, further subsidizing its small additional overhead.

It should be noted that almost all the overhead in the proposed protective scheme comes from the AC unit alone; the VA and SA are only slightly modified to accept invalidation signals from the AC. It is the AC unit which monitors the results of the VA and SA.

Additional hardware components to combat faults bring in the possibility of a fault within the new component. In this work, we assume single event upsets, i.e., only one fault could happen at any given time. This implies that there might be a fault in the VA, or SA, or the AC unit at a given time, but not in more than one of them. Hence, if there is a soft error in the VA or SA unit, it will definitely be caught by the AC unit. If, on the other hand, there is a false positive due to a fault in the AC unit, then the consequence is benign; all that the AC unit does is invalidate the allocations in the previous clock cycle. Therefore, a false positive will simply waste one clock cycle in arbitrations.

5.4.2 Routing Computation Unit Errors

A transient fault in the routing unit logic could cause a packet to be misdirected. Since the subsequent virtual channel allocation and switch allocation would be performed based on the misdirection, no data corruptin will occur. The erroneous direction, however, may be blocked, either because of a link outage (hard fault), or a network edge in various topologies (e.g., mesh). This will be caught by the VA, which maintains the state information for its adjacent routers. The VA is able to detect such erroneous behavior, because the allocator is aware of blocked links or links which are not permitted due to physical constraints (e.g., a network edge); they are either pre-programmed in the allocator's state table or they are dynamically specified through incoming state information from adjacent routers. The recovery, however, depends on whether look-ahead routing is used or not. If such a routing is employed, then the error will be caught by the VA of the following router and reported to the previous router through an appropriate NACK message. This will invalidate the erroneous decision and force the routing unit to repeat the routing process on the specific packet.

Note that the header flit is still in the previous router's retransmission buffer (as described in Section 5.3). The whole recovery process will take 3 clock cycles (one for the NACK propagation to the previous router, one for the new routing process, and one for the retransmission) in a 2-stage router. It would take 2 clock cycles in a single-stage router (one for the NACK and one for the new routing process and retransmission). To increase the efficiency in non-minimal, adaptive routing schemes, the current router may reset the invalid direction and assign a new direction without informing the previous router. If current-node routing is used (i.e., 4-stage and 3-stage routers), then the recovery phase is simpler, since the error is caught by the VA in the same router, which will inform the routing unit before the transmission occurs. This will incur a single-cycle delay for re-routing.

Misdirection to a non-blocked, functional path, however, will not be caught by the VA, since its state information will not raise an error flag (i.e., an error signal, as shown in Fig. 5.10). It could potentially cause deadlock in deterministic routing algorithms. In such algorithms, however, the error will be detected in the router that receives the misdirected flit. A NACK to the sending router would then fix the problem. The latency overhead is dependent on the number of pipeline stages (n) within the router. The delay penalty is equal to 1+n, (NACK+re-routing and retransmission). In adaptive routing schemes, the error cannot be detected. However, in such schemes, a misdirection fault is not catastrophic; it simply delays the flit traversal.

5.4.3 Switch Allocator Errors

A switch allocator error could give rise to the following four different problems, some of which would lead to packet/flit loss: (a) A soft error in the control signals of the switch allocator could prevent flits from traversing the crossbar. This case is the least problematic, since the flits will keep requesting access to the crossbar until they succeed. (b) If a data flit is mistakenly sent to a direction different from the header flit, it would cause flit/packet loss, because it would deviate from the wormhole created by its header flit. (c) A soft error could cause the allocator to direct two flits to the same output. This will lead to a corrupt flit, which will be detected by the error detection code in the next router. A NACK will be sent and the correct flits retransmitted from the retransmission buffer. Regardless of the number of pipeline stages, this error recovery process will incur two cycles (NACK+retransmission) latency overhead. (d) An error could cause the allocator to send a flit to multiple outputs (multicasting). If the flit is a data flit, the same error will occur as case (b) above. If the flit is a header flit, then multiple virtual channels will mistakenly be reserved in all the receiving routers (essentially opening multiple wormholes for the same message). Those wormholes will stay permanently reserved, thus reducing the effective buffer space in those routers.

The most challenging cases are (b) and (d). To prevent such scenarios, we propose use of the Allocation Comparator (AC) Unit, which was introduced to protect against VA errors. As shown in Fig. 5.10, the AC unit also checks for invalid SA allocations (such as multicasting) and duplicate SA allocations; upon detection of an erroneous behavior, the AC unit will invalidate the SA allocation in the previous clock cycle. In this case, the overhead involved does not depend on the number of pipeline stages of a router. In all cases, an SA error will be caught by the AC unit after the SA stage finishes. This implies that the AC unit will be operating in the same stage as crossbar traversal. Therefore, a NACK signal must be sent to all adjacent routers to ignore the previous transmission, and a new SA process will commence; this amounts for single-clock latency overhead.

We examined the impact of our proposed solutions by simulating three types of error situations. These are routing logic errors (RT-Logic), switch allocator logic errors (SA-Logic) and link errors (LINK-HBH). Each one of the cases was simulated independently by varying the error rate and measuring the number of errors corrected and energy consumption per message. Figure 5.11a illustrates the number of errors corrected by the proposed measures. Errors in the routing unit are significantly less than errors in the SA, since routing errors occur only in header flits. The SA, however, operates on every flit, and many flits often undergo multiple arbitrations before winning access to the switch. Link traversal, on the other hand, only occurs once for each flit per hop, thus the link errors detected in this experiment were less than the SA errors. Figure 5.11b depicts the energy consumed per packet under the different error schemes. As shown, link errors induce more energy overhead because of retransmissions. Nevertheless, even with retransmissions, the overhead is still minimal, thereby validating our previous assertions.

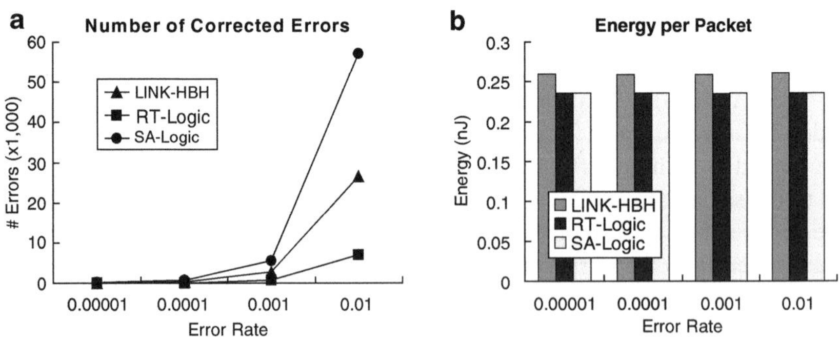

Fig. 5.11 Impact of soft-error correcting schemes. (**a**) Number of corrected errors. (**b**) Energy consumption

5.4.4 Crossbar Errors

A transient fault within the crossbar would produce single-bit upsets, not entire flits being misdirected as in the switch allocator case. Single-bit upsets are taken care of by the error detection and correction unit employed within each router, thus eliminating the problem.

5.4.5 Retransmission Buffer Errors

A single soft error in the retransmission buffer would be corrected by the error-correcting unit in the receiving router. A double (or more) error, however, would yield an endless retransmission loop since the original data itself is now corrupt. Given that a double bit-flip is highly unlikely, such a scenario can be ignored. However, a fool-proof solution would be to use duplicate retransmission buffers. This will double the buffer area and power overhead.

5.4.6 Handshaking Signal Errors

Every router has several handshaking signal lines with neighboring routers to facilitate proper functionality and synchronization. Transient faults on these lines would disrupt the operation of the network. Since the number of handshaking signal lines is small, Triple Module Redundancy (TMR) can be used, in which three lines and a voter are used to ensure protection against soft errors. There is a slight area and power overhead increase, but the area occupied by these lines is negligible compared to the area of the other router components.

5.5 Handling Hard Faults

Besides soft fault handling techniques, on-chip interconnects also require proper strategies to deal with hard (or permanent) faults in various modules, as well as the interconnect links themselves. Previous research has tackled this problem through either fault-tolerant routing algorithms [83,99,103], or by using redundant hardware for key modules in the architecture. However, a facet of permanent failures often ignored in prior studies is that they can also induce deadlocks in some routing algorithms, since a single fault can block the only allowed path for a packet. For example, in X–Y routing, if a packet directed to the X dimension encounters a fault in its path, it will be blocked there perpetually. Some prior research has successfully reflected deadlock issues in their reliability studies [82,83], but those studies target generic interconnection networks which have less constraints than on-chip interconnects. Schemes employed in off-chip and macro networks are unsuitable for on-chip implementations, because the latter impose stringent area and power restrictions. This attribute prevents the use of elaborate mechanisms in NoCs. Consequently, a simple and deadlock free fault-tolerant routing scheme is vital for on-chip reliability.

By utilizing the existing retransmission buffers of the HBH retransmission scheme introduced in Section 5.3.1 for link error recovery, we propose a simple, deadlock-free fault-tolerant routing algorithm, called Proximity-Aware (PA) Routing. The routing algorithm ensures fault-tolerance, while a newly proposed buffer management scheme resolves deadlocks.

5.5.1 Proximity-Aware (PA) Fault-Tolerant Routing Algorithm

The proposed fault-tolerant adaptive routing algorithm dynamically adapts the routing strategy based on the status of nearby nodes. The novelty in the new algorithm is two-fold: its dynamic nature allows the network to react to run-time faults in real-time. Furthermore, the algorithm uses a novel look-ahead "snooping" mechanism, which ranks possible paths as far away as two hops from the current node. This allows for the best possible path selection in the presence of faults, thus minimizing the latency penalty. In fault-free environments, packets are dynamically routed to their destination using minimal fully adaptive routing, where the distance to the destination decreases as the packet traverses the network.

In PA Routing, a node is always aware of the state of each node up to two network hops away. This health status information is received from the nodes in its direct vicinity (i.e., its four neighboring nodes). These Direct Neighbors (DN) convey information about their s status and the status of their three neighbors, as shown in Fig. 5.12. These three neighbors (i.e., the ones that are two hops away from the current node) are the Indirect Neighbors (IN) of the current node. Therefore, every node receives four different Information Sets (IS), one from each of its four DNs. These ISs are the result of the look-ahead "snooping" mechanism, and are denoted as ISE, ISN, ISW, and ISS in Fig. 5.12. Combining all four ISs, each router is now aware of the health status of both its DNs and INs.

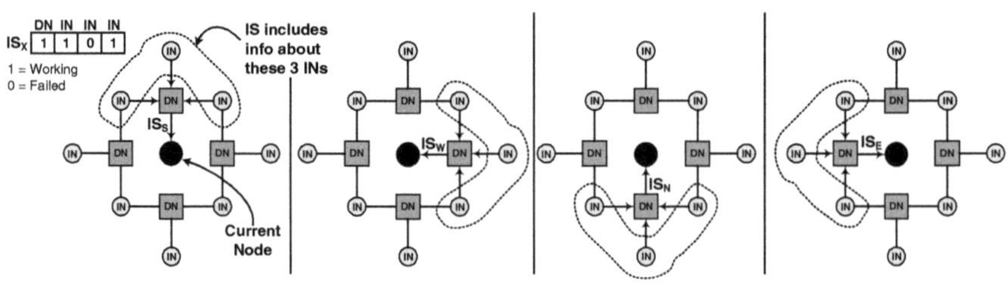

Fig. 5.12 Node health status from each direction

Based on the information from its DNs, every node can classify the nodes in its vicinity (i.e., DNs and INs) as one of four different categories – Healthy (H), Good (G), Bad (B), or Worse (W) – depending on the number of faults in each IS received; this is shown in Table 5.2. As shown in the top left of Fig. 5.12, there can never be an IS with four faults, since such a scenario would imply that the DN is itself faulty; if the DN is faulty, though, it would not be able to send the IS in the first place. Therefore, if a node is unable to receive an IS from a DN, it assumes that the DN is in a Failed (F) mode.

Using this information, each node can assign a path score to all possible paths up to two hops away, i.e., from current node to any of its INs (via one of its DNs). For example, a path scoring scheme as shown in Table 5.3 can be used (a lower path score is better). Initially, paths are pre-sorted based on the location of the destination node as compared to the location of the current node. This is illustrated in Fig. 5.13.

Table 5.2 Node status classification

# Faults	Status
0	Healthy (H)
1	Good (G)
2	Bad (B)
3	Worse (W)
4	Fail (F)

Table 5.3 Path scoring table

Score	DN	IN
0	H	H
	G	H
	B	H
1	H	G
	G	G
	B	G
2	H	B
	G	B
	B	B
3	W	*
	F	*

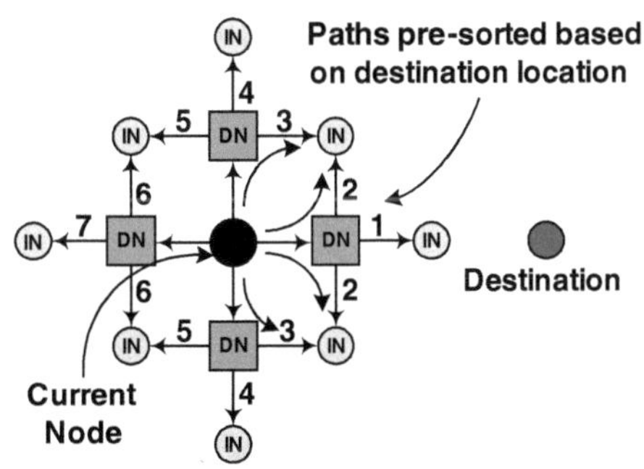

Fig. 5.13 Path pre-sorting based on destination location

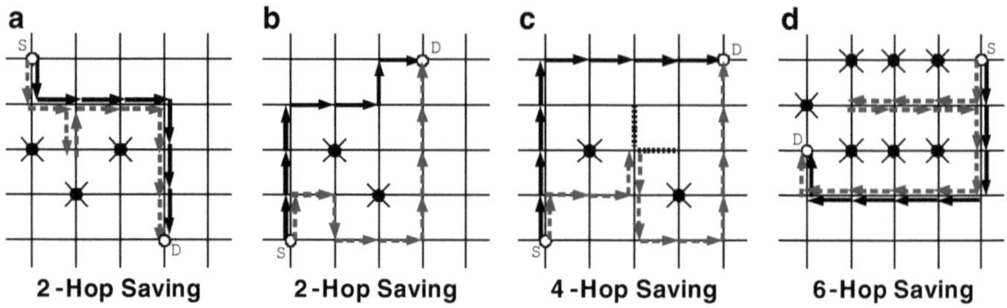

Fig. 5.14 Proximity-aware (PA) routing examples (PA: *solid arrow*, without IN consideration: *dashed arrow*). (**a**) 2-Hop saving, (**b**) 2-hop saving, (**c**) 4-hop saving, (**d**) 6-hop saving

Once the paths are pre-sorted, they are all assigned scores as described above. This score will subsequently dictate the most preferable path for a packet to take. The scoring scheme of Table 5.3 is founded on the observation that the statuses of the INs are more important than the statuses of the DNs in preventing possible backtracking of a packet. For example, if the DN is in good status (H or G), while the IN is in worse (W) status, a packet might be forced to backtrack after two hops. On the other hand, if the IN is in H status, while the DN is in bad (B) status, a packet can escape the faulty region after two hops. In other words, a healthy IN compensates for a weaker DN. Hence, the proposed PA technique with look-ahead snooping can broaden the scope of the routing algorithm by considering path traversal two hops in advance. This prevents a packet from being directed to a highly faulty area. When two paths have the same score, a decision is made based on buffer availability. It should be noted that in MESH-like topologies, where edge nodes have fewer output channels, the output directions to those missing channels will be designated as faulty channels, since a packet cannot proceed in that direction anyway.

The PA algorithm ensures that flits never go to a node in W state, where all three directions other than the input direction are blocked by faults, unless it is the destination node. This can eliminate unnecessary backtracking, as shown in Fig. 5.14a, d, as compared to a fault-tolerant algorithm without look-ahead snooping (i.e., with no IN consideration). In addition, since a router can check the status of two-hop-away nodes in advance, possible detours can be avoided that would otherwise be undetected (Fig. 5.14b, c). In general, the proposed PA routing algorithm avoids unnecessary movement of flits and can, thus, reduce overall network latency.

5.5.2 *Extension of PA Routing for Hot-Spot Avoidance*

The operational integrity of functional modules in a chip is significantly affected by excessive temperatures. Of critical importance is the avoidance of hotspots, i.e., areas in the chip where the peak temperature is extremely high. Hotspots can substantially degrade performance and adversely affect the lifetime of the chip [25].

The temperature profile is directly related to power density, i.e., dynamic power consumption over unit area. As technology scales to ever diminutive sizes, power density skyrockets. Therefore, to ensure better device reliability, thermal issues need to be addressed. Network traffic patterns in NoCs can lead to thermal issues if certain routers are overloaded. For example, NoC routers situated in the middle section of a MESH-like network topology are usually more congested than the remaining routers. Those routers are, therefore, much more likely to suffer from thermal issues. The problem can be mitigated by adopting load-balancing techniques which distribute the network workload as evenly as possible to all nodes. By balancing workloads among all nodes, peak temperatures are lower, and there is less thermal variation between routers.

The proposed PA routing algorithm can be extended to dynamically load-balance the network traffic with minimal additional logic. Existing research in temperature-aware design [75,104] has adopted either system-level distributed thermal management schemes (where no specialized hardware is required), or mechanisms that employ on-chip temperature sensors. The proposed scheme in this paper follows the former approach; it utilizes existing infrastructure, thus incurring negligible overhead.

The author's approach is based on the correlation between temperature and dynamic power consumption [104]. In turn, dynamic power consumption within a router depends on flit activity. Hence, simple monitoring of flit activity can identify possible hotspot creation.

When using the proposed PA routing algorithm, a node will not send data to blocked nodes, i.e., nodes in either W (Worse) or F (Fail) state. Therefore, if a node designates itself as blocked, other nodes will try to re-route messages via other paths to avoid the faulty region. Hence, based on average node utilization, a node can dynamically choose to switch to this artificial blocked state to reduce its workload. However, unlike actual blocked nodes, an artificially blocked node should resume normal operation as soon as the workload reduces to a desired level. Failure to do so would negatively affect the network latency in favor of load balancing. This degradation in network performance becomes more apparent in topologies where efficient traffic load balancing is inherently difficult due to blocked paths at the edge of the network (e.g., 2D-MESH topology).

This modified PA algorithm, aptly named Thermal-aware PA (TPA) algorithm, was simulated and compared to existing routing algorithms in terms of average workload distribution. The simulation assumed a MESH topology. Figure 5.15a, b illustrate the average flit activities (in flits/cycle) when using the TPA and deterministic X–Y algorithms, respectively. The XY plane represents the 8×8 MESH network, while the Z axis indicates average flit activity in flits/cycle. Clearly, the average flit activity is more smoothly distributed across the network when using the TPA algorithm, with no pronounced peaks (which could cause hotspots). Figure 5.15c depicts this trend numerically; average and maximum flit activities, as well as standard deviation (σ), are reduced in the TPA case. Reduced standard deviation implies more uniform activity distribution, without large fluctuations.

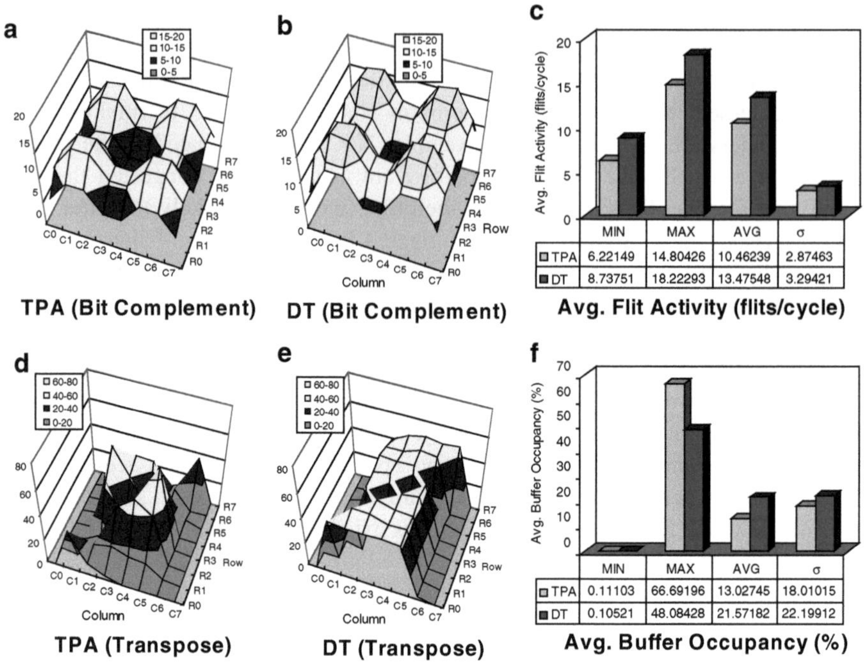

Fig. 5.15 Average load distribution. (**a**) TPA (bit complement), (**b**) DT (bit complement), (**c**) average flit activity (flits/cycle), (**d**) TPA (transpose), (**e**) DT (transpose), (**f**) average buffer occupancy (%)

Buffer occupancy may also be used as a metric for triggering the load balancing mechanism. High buffer occupancy may lead to increased blocking delay within the router [41,105]. This affects network performance. Therefore, sustained buffer occupancy over a pre-defined threshold value could invoke load balancing to alleviate the situation. This way, buffer occupancy can be more uniformly distributed over the network. Simulation results using the modified PA algorithm with this metric as compared to existing routing algorithms are illustrated in Fig. 5.15d, e. Again, the XY plane represents the MESH network, while the Z axis represents percent buffer occupancy within the routers. Similar to flit activity, the TPA algorithm redistributes traffic across the network, thereby smoothing out overall buffer occupancy. Figure 5.15f shows a significant drop in average buffer occupancy and standard deviation in the TPA case.

5.5.3 *Service-Oriented Networking (SON)*

The fault-tolerant schemes proposed so far can guarantee the transfer of a packet to its destination node. However, if the destination PE is offline due to a hard failure, the packet would circle around the destination node indefinitely. Packet loss, however, can be avoided if the packet can be redirected to a different PE that has similar or exactly the same function as that of the faulty destination node. This would allow the system to remain fully functional, even in the presence of hard faults. This destination node handover has significant importance in minimizing performance degradation in faulty environments.

Usually, a new destination is determined by the upper (software) layers in traditional networks [106]. However, if task handover can be carried out in the physical (hardware) layer instead of the software layers, the significant overhead of maintaining state and delivering signals to and from the upper layers can be eliminated. This will benefit both the software performance (since it no longer needs to worry about reliability) and overall system integration (since the hardware can dynamically reroute traffic in real-time, eliminating the need to reprogram network parameters if the PE assignment changes). Recent trends in SoCs indicate that future chips will include several PEs of the same type. Chip multiprocessors (CMP) and media processors (e.g., MPEG encoding/decoding systems) are such examples [3,4,93]. Hence, it would be beneficial to adopt such task handover schemes to further improve the on-chip reliability.

Therefore, a technique called Service-Oriented Networking (SON) is proposed, where the packet header contains a destination PE type instead of destination node address. The type of a PE is assigned based on the service that the PE offers. The router then automatically forwards a packet to the best PE of that type. Thus, even if a current destination node suddenly becomes unavailable, a packet will dynamically be rerouted to the next best PE without any disruption. By adopting this SON scheme, we can establish hardware-level packet handover to protect against node failures. To achieve this, a Direction Lookup Table (DLT) along with ultra-compact Service Information Provider (SIP) modules at each router provide the routing information to the best PE node.

5.5.4 SON – Direction Lookup Table (DLT) and Service Information Provider (SIP)

When a node detects a fault in one of its neighboring nodes, it broadcasts a message notifying about the failure. Then, all the nodes with the same PE type as that of the faulty node broadcast their locations to other nodes. Each node can, therefore, update its Direction Lookup Table (DLT), which keeps track of the best destination node for each PE type, based on a pre-defined metric. To avoid contention for access to the DLT by all input ports, a more compact version of the DLT is replicated in the form of Service Information Provider (SIP) modules, as shown in Fig. 5.16, with one SIP module per input port. Once the DLT is updated, it can then proceed to update the SIP modules. It is important to note that the DLT and SIP modules in Fig. 5.16 are not drawn to scale for clarity; they are, in fact, much smaller than the remaining router components. They contain a compact table which

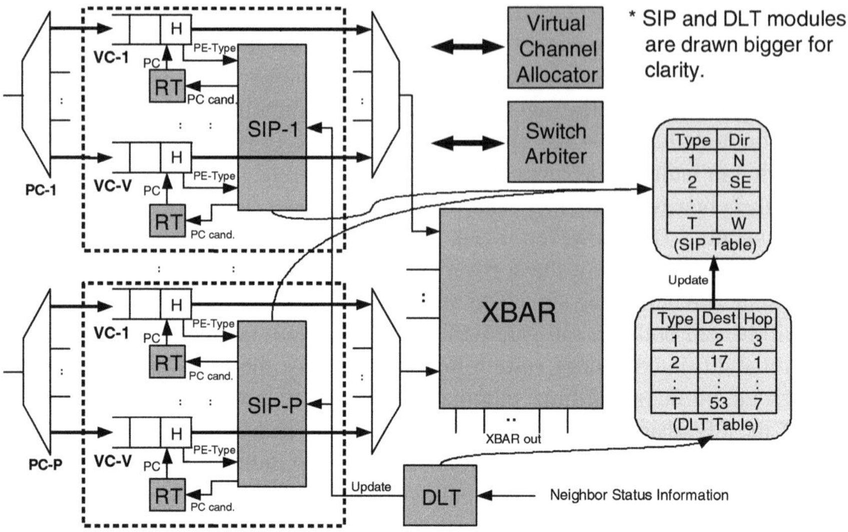

Fig. 5.16 Service-oriented networking (SON) router architecture

holds as many entries as there are different PE types. Given that the number of different PE types in a network is very small, both the DLT and SIP modules incur negligible area and power overhead.

Based on the information provided by the local SIP module, the routing unit (RT) makes a final routing decision for the packet. Since the broadcasting of faults and the DLT updates require a finite amount of time to propagate through the network, some packets may still be directed toward a faulty node, until the DLT and SIP tables are fully updated. However, this will only cause the packets to experience increased delay; it will not cause any flit loss. This delay can also be reduced by placing PEs with the same type closer to each other. In such environments, the direction to the new destination may be the same as that to the faulty node, inflicting no additional delay.

Various performance metrics can be considered when the DLT determines the best destination node for each PE type. For example, the best PE of a particular type may be the one which is closest to the current node in terms of network hops. Workload is a different metric which can also be used; the PE with the least workload is considered the best PE, if we focus on load balancing rather than network performance. Any parameter can be used as a deciding factor in SON implementations. In this paper, we used minimum distance as the handover metric. Note that workload-based handover would require periodical broadcasting of load information to each node, thereby increasing network traffic.

SON reaps more benefits from networks which include more PEs of the same type, rather than fewer PEs of many different types. In a fixed-size network, the smaller the number of PE types, the more PEs of the same type are available. In the opposite extreme, if all PEs are of different types, then a node failure would leave no alternative PEs; then, the SON architecture becomes pure overhead. However, if

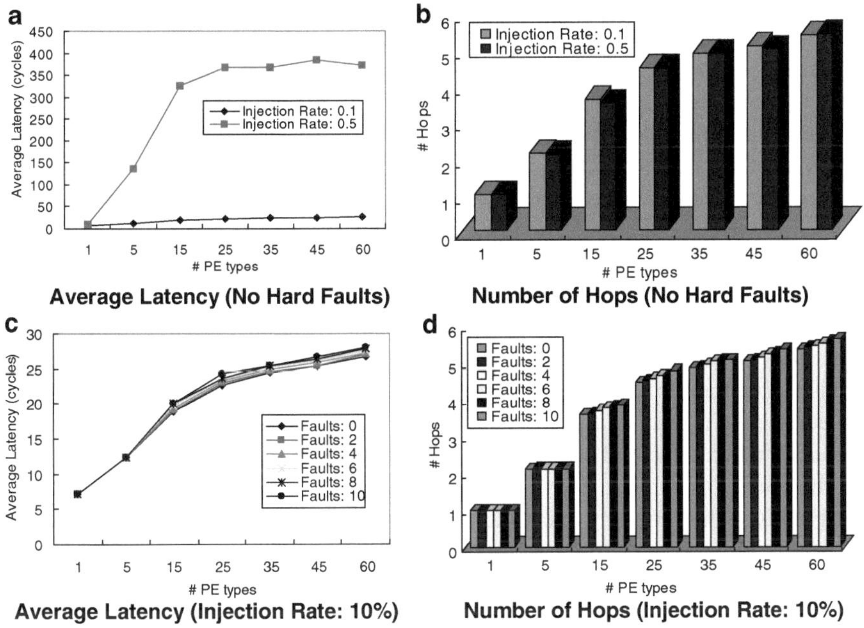

Fig. 5.17 Simulation results for SON routing. (**a**) Average latency (no hard Faults), (**b**) number of hops (no hard faults), (**c**) average latency (injection rate: 10%), (**d**) number of hops (injection rate: 10%)

redundant PEs are available, SON can efficiently utilize them under a fault scenario by minimizing the destination handover delay. Therefore, SON will be beneficial in SoC architectures with higher module redundancy, such as CMPs.

Figure 5.17a illustrates the variation in average latency as the number of PE types increases. The network is assumed to be a fixed 8×8 MESH, so an increasing number of PE types imply fewer PEs of any one type. As predicted, latency increases with the number of PE types in the network. Figure 5.17b shows the same trend, but in average number of hops; as the number of PE types increases, packets have to travel farther on average to reach a new PE of the same type. Figure 5.17c, d depict the efficiency of the proposed SON scheme. As the number of faulty PEs increases from zero to 10, the latency and average number of hops only increase slightly, highlighting the fact that the hardware handoff procedure works very efficiently in faulty environments.

5.6 Chapter Summary

In this chapter, the author presented a comprehensive plan of attack on various types of reliability hindrances in on-chip networks. Most common failure types have been tackled by proposing a series of architectural and algorithmic techniques, which work in tandem to protect the interconnect infrastructure.

A new hop-by-hop retransmission scheme was presented to combat link errors. The scheme was shown to be very efficient in terms of both latency and power even under high error rates. The retransmission buffers required by this mechanism were also used in a newly proposed deadlock recovery technique, which utilizes existing resources to break deadlocks, thus minimizing the incurred overhead. A detailed analysis of possible symptoms resulting from intra-router logic errors was also presented, along with an array of protective measures and their effectiveness in various router architectures.

Attention then shifted to hard failures. A novel Proximity-Aware (PA) routing algorithm was introduced, which employs look-ahead snooping to classify possible routing paths up to two hops away. The algorithm can dynamically adapt its routing decisions to account for run-time hard failures without any disruption in service. Unlike prior fault-tolerant routing algorithms, the proposed one can provide deadlock freedom even in faulty environments. Furthermore, the PA algorithm was extended to accommodate thermal aware routing through observation of average flit activity in the router.

Finally, a new routing methodology was presented, in the form of Service-Oriented Networking. This novel concept routes messages based on destination type, rather than destination address. This scheme takes advantage of the presence of redundant, identical processing elements in modern SoCs to provide dynamic and real-time hardware handoff in the event of hard failure.

More importantly, all the mechanisms proposed in this work kept the critical path of the NoC router intact. For on-chip networks, ultra-low latencies are an absolute necessity; thus, any reliability solution which inflicts significant burden on latency is not well suited. The schemes work in parallel with the critical components without deepening the router pipeline.

Chapter 6
On the Effects of Process Variation in Network-on-Chip Architectures [45]

Chapter 5 described how on-chip routers are increasingly susceptible to various hard and soft faults, which themselves are a natural consequence of diminutive feature sizes. In addition to these hindrances, the last couple of years have witnessed the emergence of yet another artifact of deep sub-micron technology, Process Variation (PV). PV is a consequence of manufacturing imperfections, which may lead to degraded performance and even erroneous behavior. In this chapter, the author presents the first comprehensive evaluation of NoC susceptibility to PV effects and proposes an array of architectural improvements in the form of a new router design – called ***SturdiSwitch*** – to increase resiliency to these effects. Through extensive re-engineering of critical components, SturdiSwitch provides increased immunity to PV while improving performance and increasing area and power efficiency.

6.1 Introduction and Motivation

While research in the field of NoCs has been very extensive and has addressed many pressing issues pertaining to the interconnect (e.g., performance, area, power, and fault-tolerance), a new serious impediment is currently emerging as a force to be reckoned with, which has yet to be addressed in this context: Process Variation (PV) resulting from manufacturing imperfections. The whole ideology of aggressively scaling die sizes and pursuing heavy on-chip integration is overshadowed by the aggravation of manufacturing uncertainties. PV is observed due to random effects, like Random Dopant Fluctuations (RDF) and systematic spatially-correlated effects like dose, focus and overlay variations [38]. These uncertainties impact various device characteristics, such as effective gate length, oxide thickness and transistor threshold voltages [38]. Altered device characteristics may lead to significant variations in power consumption and to timing violations. The gap between the expected and manufactured characteristics widens further as feature sizes shrink, significantly impacting the chip yield, especially in the regime of stringent timing and power budgets. Hence, process variation will instill a probabilistic flavor to circuit

C. Nicopoulos et al., *Network-on-Chip Architectures: A Holistic Design Exploration*, Lecture Notes in Electrical Engineering 45,
DOI 10.1007/978-90-481-3031-3_6, © Springer Science+Business Media B.V. 2009

design in the upcoming technology nodes of 45, 32 nm, and below. The stochastic nature of PV uncertainties requires rigorous probabilistic modeling to assess the effect of such phenomena at various levels of design integration.

Given the universally recognized importance of the interconnection network in a modern multi-core design, it is imperative that interconnect designers tackle the alarming and escalating predicament of PV. *In this chapter, the author aims to provide valuable insight to how process variation affects NoCs.* To the best of the author's knowledge, this is the first comprehensive evaluation of the effects of PV on NoC architectures. Based on sophisticated analytical models and cycle-accurate simulations, the actual impact of PV-derived artifacts on each major module of an on-chip router is first identified by propagating the variation effects on circuits to the micro-architectural level. Analysis results identify potential variations in leakage power consumption as much as 90% around the assumed nominal value, network performance degradation as much as 37%, and even up to 2.5% packet losses. After establishing the impact of PV effects on overall system behavior, three novel architectural modifications are presented to successfully combat these issues: (a) a PV-aware buffer structure, called *IntelliBuffer*, (b) a new virtual channel arbitration scheme, called *VA Compaction*, and (c) a re-interpretation of the switch allocation process in what we call *SA Folding*. These three micro-architectural innovations are combined to form *SturdiSwitch*, which offers maximum protection against PV-originating deviations. SturdiSwitch is a novel on-chip router design which provides PV-resilience without adversely affecting performance or the area and power budgets. SturdiSwitch constitutes a comprehensive plan of attack on PV-induced disruptions, and provides the NoC with a sturdy, protective blanket. *SturdiSwitch eliminates packet losses, while it manages to outperform a generic implementation by as much as 23% in the presence of PV artifacts.* By radically re-architecting the virtual channel arbitration module within the router, SturdiSwitch is, overall, more area- and power-efficient than a conventional router.

6.2 Related Work in Process Variation (PV)

Diminutive feature sizes have accentuated reliability issues in the interconnection network. Transient failures, such as radiation-induced upsets and crosstalk, and permanent faults caused by electromigration, hot carrier effects and thermal stressing are looming concerns. Several fault-tolerant techniques have been proposed to tackle both link and router failures [36,37,40,87,91,95]. However, the impact of process variation in NoC architectures remains unexplored. PV effects on inter-router links have been investigated in [107]. However, in this work, the author presents the first comprehensive analysis of the effects of PV on *all* major components of the on-chip interconnection backbone.

Manufacturing variations have become inherently unavoidable phenomena in contemporary shrinking technologies [108,109]. Such variations are quintessentially classified as die-to-die variations, namely *Inter-die* variations, and within die variations, called *Intra-die* variations. Inter-die systematic variations are primarily due to deterministic manufacturing tolerances that affect the mean value of a parameter from lot to lot, wafer to wafer and die to die [110]. Such systematic variations imply strong correlation in the properties of neighboring devices. Inter-die variations between parameters within a single die can lead to either performance degradation or failing hardware if not properly accounted for. Examples of inter-die systematic variation include channel length variation due to the length of exposure, and variations between individual metal layers used for routing. Unlike the inter-die variations, the within-die or intra-die variations may exhibit both random and systematic behavior [111]. Intra-die systematic variation comes about because of layout specific variations. These variations can be the result of semiconductor process methods or environmental differences that are seen across the design based on layout. The spatial effects are typically of deterministic nature due to inherent properties of physical processes and may be countered by techniques like Optical Proximity Correction (OPC) [112] or symmetric layout designs to reduce systematic components of Across Chip Line-width Variations (ACLV). On the other hand, random variations may be attributed to many reasons like lithography, etching, polishing and doping effects and are therefore difficult to counter. This randomness thereby causes a reason to analyze the system properties in a stochastic manner. Since all these variations impact the performance and power of the systems significantly, analyzing the design space for an optimal system prototype is quite a cumbersome process involving a stochastic scrutiny of various design options [113]. Such a statistical study enables estimation of the chip yield based upon the designer's budgets [109]. Consequently, the strong economic motivations associated with the yield warrant rigorous variation-tolerant methodologies at different levels of the design phase, ranging from circuits to architecture, as presented in [114,115]. The author's work focuses on architectural solutions in NoC-based systems to enhance the chip yield by tackling both leakage and timing impacts of variations.

6.3 The Impact of PV on NoC Architectures

6.3.1 Evaluation Platform

The cycle-accurate NoC simulator from Chapter 5 was used in order to conduct a detailed evaluation of the on-chip network architectures discussed in this paper. The simulator operates at the granularity of individual architectural components, accurately emulating the major hardware components. The simulation test-bench models both the router components and the inter-router links. All experiments assumed a 64-node Mesh network (8×8), the use of deterministic (XY) routing under uniform random traffic, and two-stage routers (RC, VA and SA in stage one, and XBAR/Link in stage two). Each router has 5 input/output ports (i.e., $P=5$), 4 VCs/

port (i.e., $v=4$), 4 flits/VC (i.e., $k=4$, and $5\times4\times4=80$ buffer slots per router). One packet consists of four 128-bit flits. The simulator keeps injecting messages into the network until 300,000 messages (including 100,000 warm-up messages) are ejected. XBAR/Link traversal was assumed to complete within one clock cycle [48].

Furthermore, the modules comprising a typical NoC router (as described in Chapter 2) were implemented in structural Register-Transfer Level (RTL) Verilog HDL and then synthesized in 90 nm technology using Synopsys Design Compiler and commercial-grade TSMC libraries. The resulting designs operate at a supply voltage of 1 V and a clock speed of 1 GHz. Both dynamic and leakage power estimates were extracted from the synthesized router implementation. The numbers were then back-annotated into the cycle-accurate network simulator and used for power profiling, based on the activity factor of each component.

The delay models for PV analysis were developed in MathWorks MATLAB and integrated with Synopsys Design Compiler. Our model is depicted in Fig. 6.1. A synthesized gate-level netlist of the provided HDL implementation was used by a MATLAB function to generate variable gate delays. Such delays are generated based on both the systematic and random effects contributing to process variations. While, the random delays are modeled as a normal distribution varying about the mean gate delays with a deviation of 5%, the systematic variations need a model capturing spatial correlations. The hierarchical correlation model presented in [111] was used to model such variations. As illustrated in Fig. 6.1, a given design floorplan is tiled into a number of rectangles of different sizes. The different sizes aid in capturing spatial correlation of varying extant. Any systematically varying property, P, of a component in a tile is modeled as a sum of random variables accrued

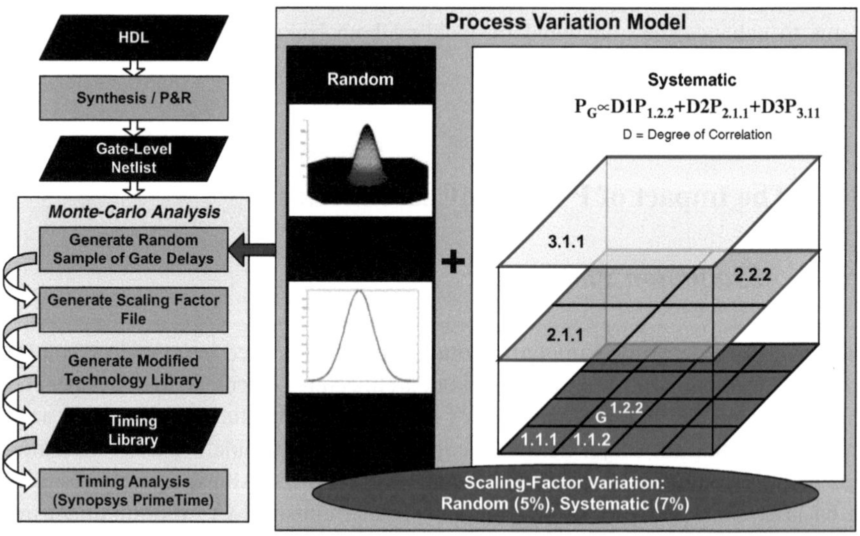

Fig. 6.1 Methodology used to model process variation effects in the NoC router architecture

over tiles at all sizes to which the component belongs (e.g., for point G, tiles numbered 1.2.2, 2.1.1 and 3.1.1 play a determining role). This model captures the fact that the property of any gate is dependent upon the neighboring gates in the same tile at different levels, where the topmost layer essentially captures the randomness associated with the whole chip (inter-die variations). The degree of correlation is defined by the correlation coefficients of neighboring tile properties in the lowermost level.

The modified gate delays obtained from this model were then incorporated back into the TSMC libraries. Finally, the timing of the designs was re-estimated using Synopsys PrimeTime. The statistical nature of the MATLAB code (which generates modified gate libraries) integrated with a Monte Carlo-based evaluation enables comprehensive statistical timing analysis of any HDL code provided. Variation analysis of leakage power in buffers requires precise power estimation, which is best performed using an HSPICE-based study. Therefore, a 128-bit buffer (i.e., one buffer slot) was designed and simulated for statistically-varying device parameters, and the leakage power variations were observed. The leakage power simulations employed the PTM device models [116] for both 90 nm and 32 nm technology nodes, since leakage power is expected to dominate dynamic power in deep submicron feature sizes (45, 32 nm, etc.). The details of the individual variation models are discussed in Section 6.3.2.

Before embarking on our PV exploration, it is important to identify the contribution of each major router stage to the overall critical path delay. This information will be instrumental in guiding our proposed modifications to improve PV-resilience *without adversely affecting performance*. Hence, detailed timing analysis using Synopsys PrimeTime and the synthesized designs yielded the critical paths of the major router components. These numbers are shown in Fig. 6.2, which depicts the

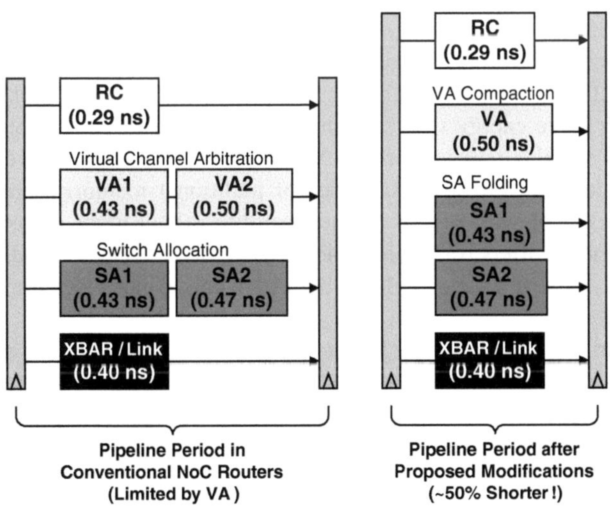

Fig. 6.2 Critical path delays of all router stages (**LEFT**: generic, **RIGHT**: SturdiSwitch)

critical path delays of all intra-router stages for a conventional router (left) and the proposed SturdiSwitch router (right). SturdiSwitch will be described in detail later on. Note that the numbers of Fig. 6.2 do not include clock and latch overheads; they are the critical path lengths of the purely combinational logic.

Observation 1: *Clearly, the pipeline clock period in a conventional router is determined by the VA and SA stages, with the RC and XBAR/Link stages exhibiting significant slack.*

6.3.2 PV Effects on Router Components

As indicated in **Error! Reference source not found.**, the NoC router is mainly composed of the input buffers storing the flits, the arbiters governing virtual channel and switch arbitration, and the crossbar/links for the transmission of packets to adjacent routers. Hence, to study the impact of process variation on NoC architectures, we perform analysis on each individual component. The variation study is conducted at two different granularities, for each of the components: (1) microscopic (i.e., variation impact at the circuit level) and (2) macroscopic (i.e., the architectural implications of PV on component/module operation). We demonstrate how different anomalies introduced due to variations in the circuits impact the performance, power consumption and even functionality of the entire system. The impact on individual components is discussed further in the following subsections.

6.3.2.1 Input Buffers

Among the different components comprising an on-chip router, buffers are the largest leakage power consumers; about 64% of the total router leakage power is attributed to the buffers [50]. In fact, it has been observed that storing a packet in a buffer consumes far more energy than transmitting the packet [51]. The buffers can be implemented as either registers or SRAM/DRAM memory [69]. However, to avoid the address decoding/encoding latencies of traditional memories and the access latencies associated with global bitlines/wordlines, it is more reasonable to use small registers as buffers [69]; this choice is justifiable given the small buffer sizes used in NoC routers. Thus, the NoC buffers in this paper were implemented as registers.

It is important to note that the buffers are accessed only when the packet enters or exits the router. Considering control latencies and wait times in the system, the buffers typically tend to store packets for considerable periods of time, especially in highly-congested areas of the network. Moreover, the nature of the routing algorithm employed may lead to unbalanced buffer utilization, which in turn causes long idle periods in a skewed manner. As a consequence of such observations, the average active and standby leakage power consumption of the buffers become quite significant when considering the design of NoC routers. On the other hand, since buffers are accessed only during the XBAR/Link stage of the router's operation, the impact of buffer read/write delay variations is nullified by the slack associated with this stage (see Fig. 6.2). This realization motivates us to focus our attention only on the power impact of process variation on the buffers, and particularly leakage power due to its dominance.

As mentioned earlier, the variations in leakage power for both the 90 and 32 nm technology nodes are analyzed. A 128-bit register file was used to test the variation impact on the leakage power consumption due to different parametric variations. The model for variations used in this work incorporated both systematic correlations and random variation effects. The model assumed parametric variations with total standard deviations of 15% on effective gate length (L_{eff}) and threshold voltage (V_{th}), distributed evenly amongst inter/intra die variations using the methodology presented in [111]. Monte Carlo statistical analysis using the normal distributions of the parameters was performed in HSPICE.

Figure 6.3 demonstrates the variations in the total leakage power consumption of a single 128-bit buffer slot at the 90 nm technology node. As expected, the buffer leakage power follows a log-normal distribution due to the exponential dependence of leakage power consumption on the process parameters. The minimum and the maximum variations differ by close to 4×, as observable from the figure. The leakage distribution, when fitted to a normal distribution, demonstrated a 90% variance around the nominal leakage power consumption. Such leakage power distribution along with the utilization of the VC buffers under various load rates (obtained from our NoC simulator) will be used in Section 6.4.1 to determine the variation in the total system leakage power consumption.

Observation 2: *The impact of PV on leakage power consumption in the NoC buffers is very significant: up to 90% variation around the nominal value has been demonstrated.* As leakage power becomes increasingly dominant with technology

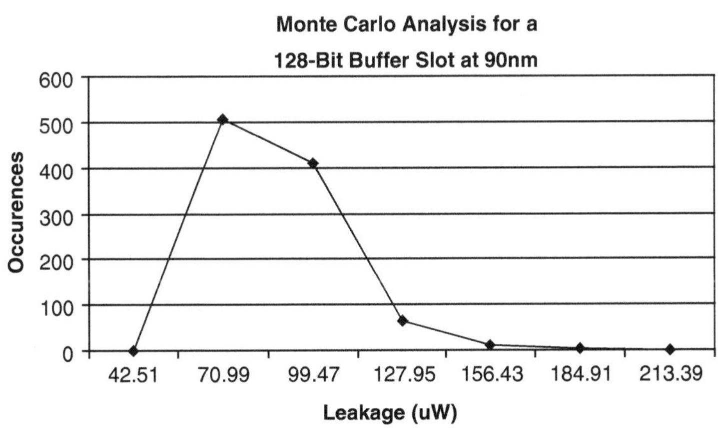

Fig. 6.3 Variations in leakage power consumption of a single buffer slot (128 bits)

scaling, PV-induced artifacts will become even more pronounced. Therefore, *designing PV-resilient buffers is essential in curbing this spiraling trend.*

6.3.2.2 Virtual Channel Arbitration (VA) and Switch Allocation (SA)

The critical combinational delay of the NoC router is typically governed by the arbiter delays in the VA and SA stages. Consequently, the criticality of these arbiters in determining the maximum clock frequency of the entire router calls for rigorous timing analysis of these components. Obviously, delay variations in the arbiters may lead to pipeline timing violations. We first demonstrate how variations in the arbiter's logic can lead to timing discrepancies, followed by a detailed study of the impact of such delay anomalies to the overall router operation.

A q:1 arbiter is a q-input arbiter producing one winner, based on a selection policy (e.g., round-robin). The output of the arbiter is typically one-hot encoded, with the active line indicating the arbitration winner.

An unexpected side-effect of the two-stage VA and SA processes is increased vulnerability to PV-induced delay effects. Because of process variation, the first stage arbiters can have different propagation delays, even though they are logically identical. This delay variation will manifest itself in differing arrival times at the inputs of the second stage arbiters, causing glitches at their outputs. Such glitches could be latched in the output registers, potentially giving rise to erroneous behavior.

The investigation of this issue aims to determine (1) whether such glitches are frequent enough to warrant a safeguard, and (2) the effect of these glitches on the router's operation (i.e., whether they are malignant or not). The impact of PV on an arbiter was first studied by using the Monte Carlo-based statistical evaluation platform described earlier. The individual gate-delay variation modeling employed in this work is similar to the one presented in [111]. The variations in delay are modeled as a sum of the *Random* and *Systematic* variations. Systematic variations are modeled by considering path-based spatial correlation using the three-layer hierarchical approach presented in [111]. Figure 6.4 demonstrates the result of a Monte Carlo analysis for the timing of a 20:1 round-robin arbiter (the biggest arbiter used in our simulation environment) for different correlation coefficients (value of 0.5 indicates half the chip is correlated in the lower-most level of the three-level correlation model [111]). The narrowing of the normal curves with increasing correlations is attributed to the reduction in the systematic variation contribution, which makes the variations tend toward deviations mainly due to random effects. The graph clearly indicates that the timing variation is significant enough to potentially cause glitches in cascaded arbiter configurations. For a correlation of 0.3, the 3σ variation in timing is around 13%.

The most significant outcome of this exploration is the discovery that *glitches in the VA process will, in fact, lead to erroneous behavior (and not simply performance degradation)*. Our analysis indicates that propagation delay variability resulting from the two cascaded arbitration stages may lead to *three different kinds of errors.*

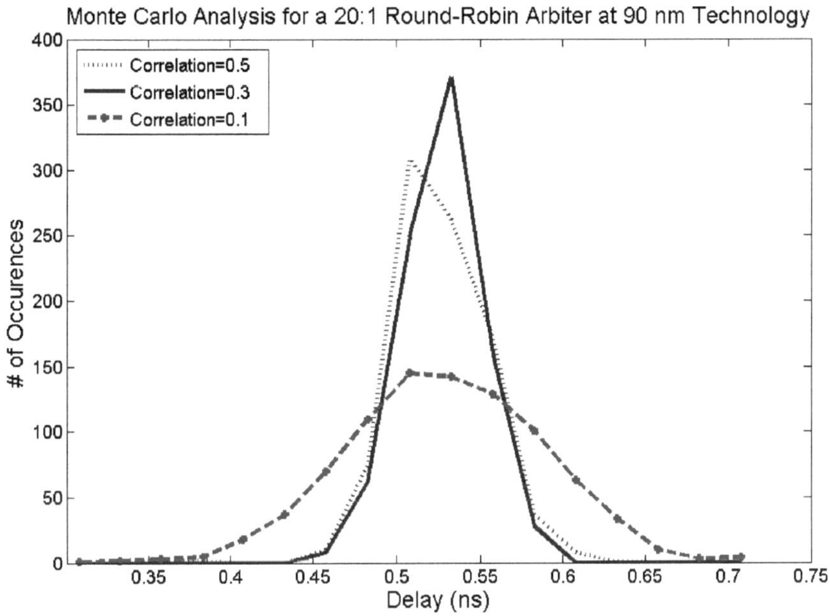

Fig. 6.4 Monte Carlo timing analysis for a single 20:1 round-robin arbiter

(1) No Grant: Delayed arbitration may lead to the pipeline latch missing the output grant signal from the arbiter. This is a relatively benign error, because it only leads to performance degradation (missed clock cycles). In the VA case, the degradation is less pronounced, because VA is a per-packet operation. In the SA case, though, the effect is much greater, since it leads to idle output ports (i.e., wasted throughput) on a per-flit basis.
(2) Priority Violations: A fast-arriving signal is latched as the granted signal from the arbiter, even though the priority for transfer is with a slow-arriving signal. This error is benign, since the slow-arriving signal will eventually win arbitration in the following clock cycle, and, in terms of overall operation, no cycles are wasted.
(3) Double-Grants: A spurious glitch at the output of a second-stage arbiter (VA2/SA2) because of PV-induced delays in the first-stage arbiters (VA1/SA1) may end up getting latched and creating a double grant. This primarily happens due to the late fall of the false grant and simultaneous rise of the actual grant (a consequence of varying input arrival times), as illustrated in Fig. 6.5. A double-grant at the output of a second-stage VA arbiter implies that the same output VC is erroneously granted to two different packets. The severity of such faults is critical, since they lead to packet loss. Two packets sharing the same VC will result in packet mixing. In wormhole routers (which is the norm in on-chip communication), packet mixing implies that the flits of one packet will follow the wrong wormhole behind the other packet's header flit. Similarly, a double grant at the output of a second-stage SA arbiter would send two waiting flits to

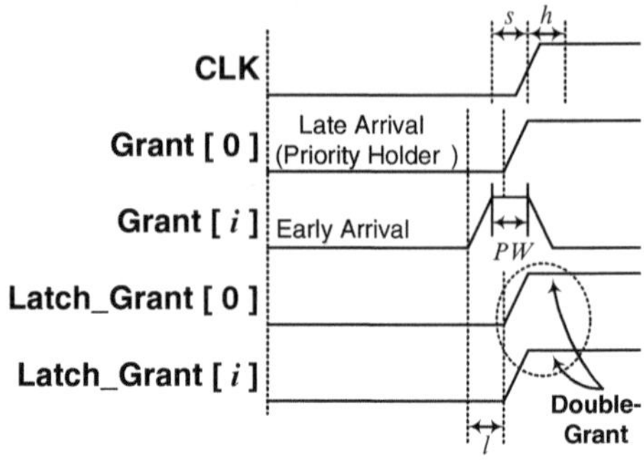

Fig. 6.5 Double-grant arbiter output resulting from non-synchronized inputs

the same output physical link at the same time. This, of course, will corrupt the data on the output channel and result in flit loss. If retransmission provisions are present in the router, communication will resume after retransmission of both lost flits. In this case, the consequence will be reduced network throughput. Many on-chip routers, however, do not employ retransmission schemes, because such schemes entail additional buffer space, which is undesired in resource-constrained environments [37]. Instead, they rely on forward error correction which guards against a very limited number of bit errors per flit. In such cases, the router would be unable to deal with flit-mixing, which involves multiple bit errors.

Therefore, ***double-grants in the VA and SA units are of critical importance, since they lead to flit/packet losses***. The frequency of these upsets will continue to grow as technology scales.

The probability of each of the above three events occurring is demonstrated statistically, beginning with double-grants and extending the analysis to no-grants and priority violations. The probability of a double-grant is completely determined by the width and position of the glitch. A statistical formulation for the event of a double-grant is provided and the probability of such an event is estimated. The probability of a grant getting latched, P_{GL}, is shown in Eq. (6.1). The equation is derived based on a grant satisfying the setup time (s) condition of the latch operating at a clock period C. Equation (6.2) depicts the probability P_{PWL} of a variation-induced glitch of width PW getting latched by a flip-flop (i.e., register), similar to the expression presented in [117] for soft errors. Parameter P_L in the equation indicates the combinational logic delay while h is the hold time of the latch. The denominator of the equation represents the period of possible occurrence of the glitch while the numerator captures the condition of a glitch of pulse width PW getting latched. Equation (6.4) shows a way to estimate the probability of generating a glitch of pulse-width PW in a stacked-arbiter-based system. This equation is derived on the basis of a maximum gap in the arrival times, l, of different input signals to the arbiter (coming from a preceding, cascaded arbiter). The arrival time difference is computed as the difference between two similar normal distributions

(A (μ, σ)), as shown in Eq. (6.3). In Eq. (6.4), P_{rk} represents the probability that an input signal, *k*, having priority to win (based on the arbiter selection policy) makes a request to the arbiter. P_{pk} is the probability that input signal *k* has the priority to win. P_{ri} is the probability that another input signal, *i*, also makes a request to the arbiter. Note that the pulse width *PW* is equal to the difference in arrival times of the input signals. The parameter P_p is the probability of an input having priority in the arbiter, and P_r is the probability of an input making a request to the arbiter. Finally, the probability of a double-grant, P_{DG}, is estimated as a continuous integral of conditional probability, expressed as the triple product of a glitch of a given width occurring (P_{PW}), the probability of it being latched (P_{PWL}), and the probability of the other grant with priority getting latched (P_{GL}), as shown in Eq. (6.5). Similarly, distributions for the probability of a no-grant, P_{NG}, and priority violations, P_{PV}, are given in Eqs. (6.6) and (6.7), respectively.

$$P_{GL} = 1 - \frac{s}{C} \tag{6.1}$$

$$P_{PWL} = \frac{PW - (s+h)}{C - P_L} \tag{6.2}$$

$$l = A(\mu,\sigma) \Rightarrow (l_1 - l_2) = A\left(0, \sqrt{2}\sigma\right) \tag{6.3}$$

$$\begin{aligned} P_{PW} &= P(W = PW) \\ &= P_{rk} * P_{pk} * \sum_{i \neq k} \left[P_{ri} * (1 - P_{pk}) * P\left(A_i\left(0, \sqrt{2}\sigma\right) = PW\right) \right] \\ &= (N-1) * P_r^2 * (1 - P_p) * P_p \\ &\quad * P\left(A\left(0, \sqrt{2}\sigma\right) = PW\right) \end{aligned} \tag{6.4}$$

$$P_{DG} = \int_{PW=(s+h)}^{\infty} (P_{PW} * P_{PWL}) * P_{GL} \tag{6.5}$$

$$P_{NG} = \left[1 - \int_{PW=(s+h)}^{\infty} (P_{PW} * P_{PWL}) \right] * (1 - P_{GL}) \tag{6.6}$$

$$P_{PV} = \int_{PW=(s+h)}^{\infty} (P_{PW} * P_{PWL}) * (1 - P_{GL}) \tag{6.7}$$

To test the impact of such anomalous behavior on the whole system, the values of the different probabilities were obtained and back-annotated to the NoC simulator. The distribution of the delays for an arbiter obtained from the analysis shown earlier was used for different arbiters operating at 1 GHz. The probability P_p is approximated to *1/(NUMBER OF INPUTS)* in these simulations, which is a reasonable assumption for the fair round-robin arbiters assumed in this work. Parameter P_r is a function of the network traffic patterns, and is assumed to be the average inter-arrival flit time in the router. The values of P_{DG}, P_{PV} and P_{NG} were calculated using the equations above as 0.02, 0.00036 and 0.089, respectively, for a 20:1 round-robin arbiter. This indicates an alarmingly high probability (as much as 2%) of double-grants, which translates directly to packet losses. Figure 6.6 depicts the percentage of packets lost as a result of double-grants, for different load rates. Double-grants can cause up to 2.5% packet loss. The 8.9% of wasted cycles (for the no-grant case) was simulated in our cycle-accurate network simulator to assess performance degradation. The average increase in network latency is plotted in Fig. 6.7 for different load rates. Figure 6.7 illustrates the effect on performance resulting from no-grants in the VA and SA arbiters separately to highlight the fact that no-grants in the SA unit have a more pronounced impact (around 37% as opposed to the VA's 9%) on network latency, as previously explained. The "Ideal" column in Fig. 6.7 represents an environment with no PV effects.

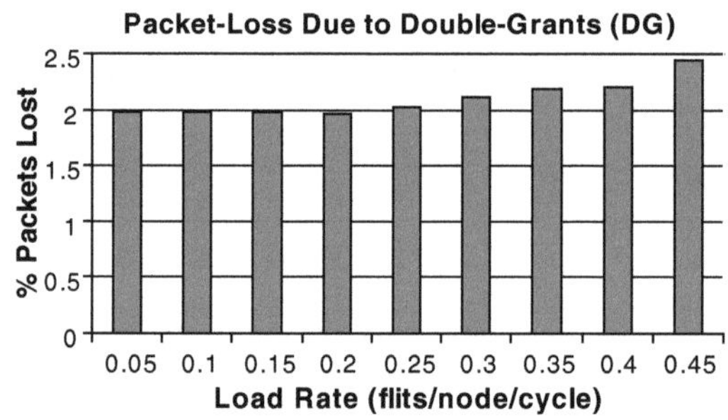

Fig. 6.6 Impact of double-grants on network reliability

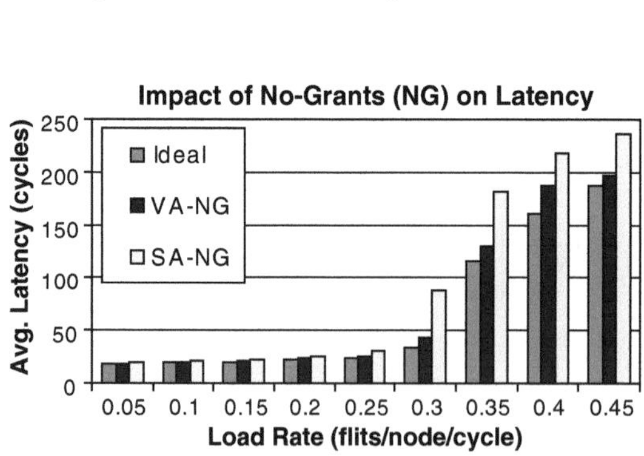

Fig. 6.7 Impact of no-grants (i.e., clock skips) on network performance

The occurrence of priority violations is quite infrequent due to the statistically low probability of such an event.

Observation 3: *The impact of PV on the arbiters of the VA and SA modules is serious enough to warrant the addition of protection logic*. The effects range from considerable performance degradation (up to 37%) to erroneous behavior (up to 2.5% packet losses). If ignored, PV will hamper the overall operation of the on-chip network. Our timing experiments indicate that *double-grants can be completely eliminated if the inputs to all arbiters emanate from synchronous sources* (i.e., registers). In fact, *synchronous sources also eliminate priority violation errors*, leaving only no-grants as the sole PV-related effect in the VA/SA stages (because of possible delays within the arbiters themselves). *This result is of profound significance and clearly defines our design goals hereafter: (1) synchronize the inputs to all arbiters, and (2) minimize the impact of no-grants.*

6.3.2.3 Crossbar (XBAR)/Links

Process variation in the XBAR and links will translate primarily into timing/delay variations in the different output links. Because NoC inter-router links are implemented in the top (global) metal layer, their delay is minimal. Our analysis using the PTM model [116] indicates that even for distances as large as 5 mm, the link delay is one order of magnitude smaller than the intra-router delay. The work in [48] also verifies that link delay is substantially smaller than the router delay. Figure 6.2 and Observation 1 highlight the fact that the XBAR/Link stage enjoys a generous slack. Furthermore, the significant slack of the XBAR itself is also validated in [49]. Therefore, *delay anomalies in the XBAR and links will be absorbed by the slack and their effects rendered inconsequential to the router operation*. This assertion agrees with [107], which presented an analysis of the timing variations in NoC links due to PV and concluded that systems having the link stage operating within a slack of 20% or more will not be impacted. Since typical NoC routers have a slack of more than 20% in the XBAR/Link stage the impact of delay variations on this stage is negligible.

6.4 The Proposed SturdiSwitch Architecture

The results of Section 6.3 clearly indicate the need for PV-aware router designs. Toward this extent, *a novel PV-resilient router architecture* is proposed, aptly named ***SturdiSwitch***. SturdiSwitch encompasses *three fundamental micro-architectural innovations*, which collectively address all three stern issues identified in the previous section: (1) buffer leakage power, (2) vulnerabilities in the VA unit, and (3) vulnerabilities in the SA unit. The three key components comprising the SturdiSwitch design are presented in detail below.

6.4.1 *IntelliBuffer: A Leakage-Aware Elastic Buffer Structure*

As previously explained, each router input port has a dedicated k-flit FIFO buffer for each of its v virtual channels. The ultra-low latency demands of on-chip communication dictate the use of a parallel FIFO implementation, which does *not* require a flit to traverse all slots in a pipelined manner before exiting the buffer. To achieve this feat, the buffer relies on read/write pointers to maintain the FIFO order. The pointer logic is, in fact, very simple: both read and write pointers keep shifting by at most a single slot position per clock cycle in the same direction. This behavior can be fully captured by two wrap-around shift registers, each of depth k, as illustrated in Fig. 6.8 for $k=4$. An entry of logic one in the shift registers indicates the read/write pointer locations. Hence, the pointer logic ensures predictable behavior in buffer usage; all slots are utilized in strict sequential order, with no "jumps" or random accesses.

It is precisely this predictable, sequential behavior of FIFO buffers that renders them susceptible to process-variation induced leakage anomalies. "Leaky" slots cannot be differentiated from other slots, because of the rigid sequential access pattern which cannot bypass problematic areas in the buffer. Thus, on average, leaky slots are used as much as the other slots.

To tackle this issue, a powerful modification to the pointer logic of the FIFO buffers is proposed to minimize leakage-power consumption. The resulting implementation is called ***IntelliBuffer***, because the buffer can now intelligently adapt its operation to utilize the less leaky slots as much as possible, while marginalizing the use of leaky ones.

IntelliBuffer differs from a conventional FIFO buffer in two fundamental ways. First, its read/write pointers do NOT blindly follow a sequential order. Instead, the write pointer always tries to direct incoming flits to the least leaky slots, based on leakage pre-classification of all slots in the buffer (i.e., the slots are classified in advance, based on their leakage characteristics). Second, all unused slots are supply-gated using sleep transistors to minimize leakage power consumption. To avoid

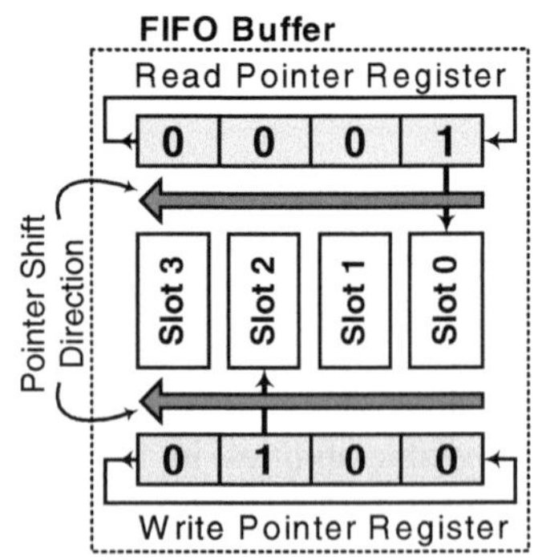

Fig. 6.8 A four-slot FIFO buffer and its control logic (pointers)

performance penalties related to wake-up latency, the write pointer is utilized to power up the next empty slot as soon as the current slot starts being written to [118]. This look-ahead power-up policy ensures that an empty slot is always active in anticipation of the next incoming flit.

The combination of these two attributes creates a leakage-aware elastic buffer, with active slots "expanding" and "contracting" based on usage (i.e., incoming traffic). Most importantly, though, this expansion/contraction is not done arbitrarily; *the leakiest slots are always the last to be used, ensuring that, under light traffic load, they are almost always switched off.*

The IntelliBuffer architecture is shown in Fig. 6.9. Additions to the generic control logic are kept to a minimum: the slot IDs are now pre-classified in order of "leakiness" in the very compact Leakage Classification Register (LCR). The write pointer logic always chooses the first available slot, which is highest in the leakiness list (i.e., least leaky). As slots are being written to, the slot IDs are written into a small FIFO buffer (the Slot ID List). The head of this list forms the read pointer of the main IntelliBuffer flit FIFO, which holds all incoming flits. Conventional FIFO buffers are dominated by the datapath delay (i.e., reading and writing large flits into the slots). Thus, the pointer logic latency is masked by the datapath latency. Based on this realization, the additional control logic of IntelliBuffer was architected to operate in parallel with the buffer's datapath; the control logic latency is completely hidden by the datapath latency. The synthesized results validate the assertion that IntelliBuffer is, in fact, as fast as a conventional buffer structure. In 90 nm technology, the critical path delay of the new control logic is 0.24 ns, which is smaller than the buffer datapath delay of 0.35 ns. Additionally, because the overall buffer structure is dominated by the large flit slots (the Flit FIFO in Fig. 6.9), the additions to the control logic inflict a negligible 4% area and power overhead over the conventional FIFO buffer. The minor

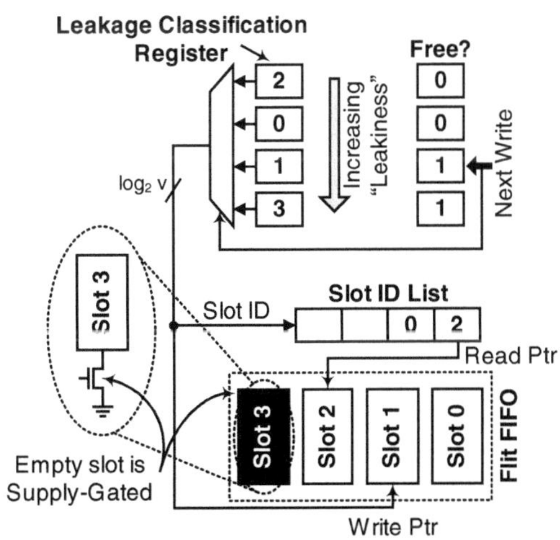

Fig. 6.9 The proposed IntelliBuffer architecture (one FIFO buffer shown)

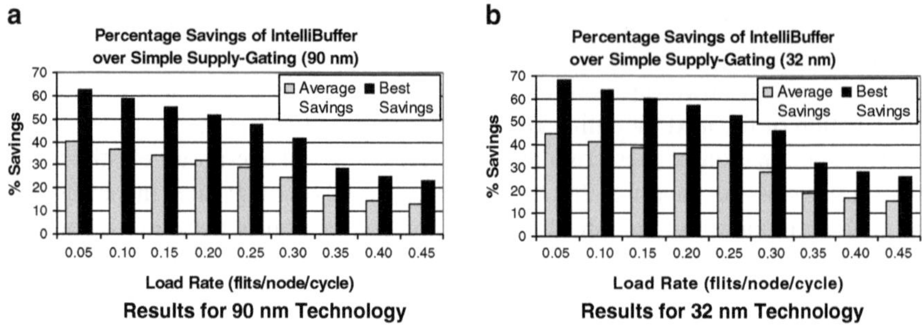

Fig. 6.10 IntelliBuffer leakage power savings. (**a**) Results for 90 nm technology. (**b**) Results for 32 nm technology

power increase is, of course, compensated by the much larger savings extracted from the main buffer slots. Figure 6.10 illustrates the leakage power savings generated by the IntelliBuffer architecture, as opposed to a structure which simply supply-gates empty buffer slots [118], for 90 nm and 32 nm technologies. IntelliBuffer achieves substantial benefits of around 24%, on average, at 90 nm, and savings of around 28% at 32 nm. Note that the parametric variations at 32 nm are expected to be even higher, which would accentuate the savings due to IntelliBuffer. Furthermore, even though the difference in *percentage* savings between 90 nm and 32 nm is small, the *absolute* savings are much greater, since the 32 nm node is significantly leakier.

The Leakage Classification Register (LCR) is populated off-line according to measurements collected during a built-in self-test (BIST) sequence which probes each individual buffer slot of the router, as described in [119]. BIST techniques are beyond the scope of this paper, so this exploration/evaluation assumes that the LCR is filled offline and ready to use at router startup.

6.4.2 VA Compaction Mechanism

Section 6.3.2.2 identified the two cascaded stages of arbitration in traditional VA units (VA1 and VA2 in Fig. 6.10 (a)) as the primary culprits in the creation of glitches which may be latched at the output of the VA stage. Our goal is to modify the VA logic so as to eliminate the occurrence of malignant double-grants, which lead to packet/flit losses, and minimize the impact of no-grants on system performance. A somewhat naïve solution to this predicament is to simply increase the clock period to allow the glitches to subside. However, such an approach would be detrimental to the overall router performance, since the VA stage is already the slowest of all the pipeline stages (see Fig. 6.2); an increase in its critical path would inevitably lower the operating frequency of the entire router.

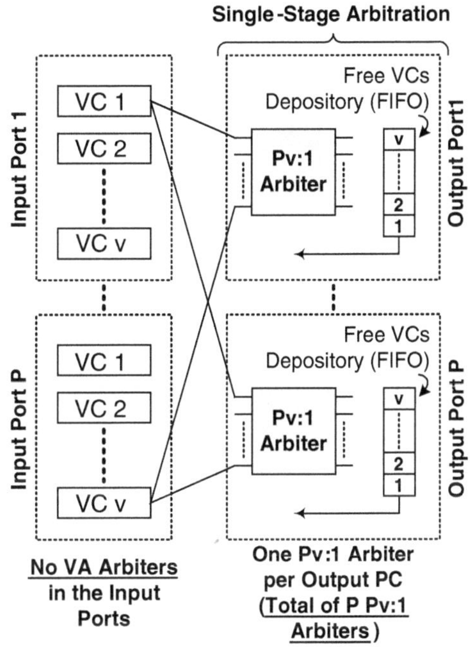

Fig. 6.11 The VA Compaction architecture

Hence, the author proposes a radical rethink of the whole VA process which will result in a PV-resilient configuration. The proposed solution, called ***VA Compaction***, eliminates the first arbitration stage altogether to ensure synchronized arrival times at the inputs of the second-stage (global) arbiters. Furthermore, it even reduces the number of second-stage arbiters from Pv to P, without affecting performance. An architectural overview of VA Compaction is shown in Fig. 6.11 (the conventional VA unit is shown in Fig. 6.10 (a)).

The purpose of the first-stage arbiters in a conventional VA unit is to decide which output VC the packet should request for a particular output port. However, through VA Compaction, this task can be removed from the VA process by employing a free-VC list. Each packet can now directly request a free VC from a particular port, and once it wins arbitration, a free VC will be granted from the free-VC list. This modification avoids the need to arbitrate between multiple free VCs. While the idea of using a free-VC list for the output VCs was used in [48], VA Compaction introduces another key improvement: it reduces the number of second-stage arbiters from Pv to P. This reduction stems from the fact that in VA Compaction, only one arbiter is required per output port. The arbiter takes in requests from all packets which require a VC in that particular output port. Instead, in all existing VA implementations, one arbiter is required for *each output VC*. Careful examination of the operation of an on-chip router, however, reveals that such overhead is unnecessary and yields absolutely no performance benefits.

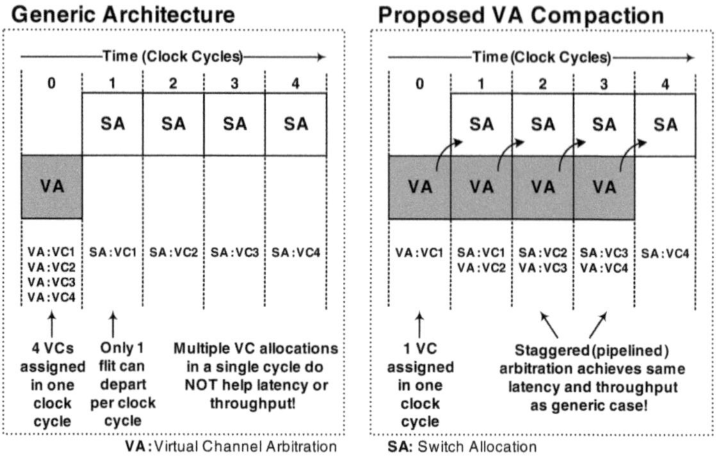

Fig. 6.12 The staggered arbitration mechanism employed by the VA compaction architecture

The reason is the existence of an input DEMUX in each input port (see Fig. 6.9). The DEMUX creates a bottleneck by limiting the number of flits that can arrive at (or leave from) a port in a single clock cycle to *one*. This means that even if the VA unit could grant multiple VCs in the same output port per cycle, only one flit can leave for that port at a time. Based on this insight, VA Compaction requires only one arbiter for each output port. Thus, by exploiting a staggered, pipelined arbitration mechanism, as shown in Fig. 6.12, VA Compaction can achieve the same throughput, even if it only grants a single VC per port per clock cycle.

In fact, VA Compaction also yields significant improvements in performance, because it dispenses free VCs in a much more efficient manner. In a two-stage VA process, the first-stage (local) arbiters function independently and are oblivious to each other's operation. Thus, there is a high probability that two packets destined for the same output port will decide to compete for the same output VC, even though there is more than one available output VC. In such a scenario, the two packets will clash in the second (global) arbitration stage, and only one will win access to the requested output VC. The loser will be forced to re-arbitrate all over again from stage 1. This pattern may be repeated more than once under high traffic loads. By contrast, such inefficiencies are absent in the proposed VA Compaction scheme, due to the presence of a centralized free-VC list in each output port. Figure 6.13 shows that *routers equipped with VA Compaction can achieve a marked improvement of around 30% in average network latency for higher load rates, as compared to a generic VA implementation* ("VA Comp" vs "Generic" curves). Furthermore, routers with VA Compaction start saturating at much higher load rates, as indicated by the graph.

The benefits afforded to the NoC router by VA Compaction are multi-faceted and of profound significance: (1) the logic depth of the VA unit is cut in half (first-stage arbitration is removed). Given that VC arbitration is the slowest module

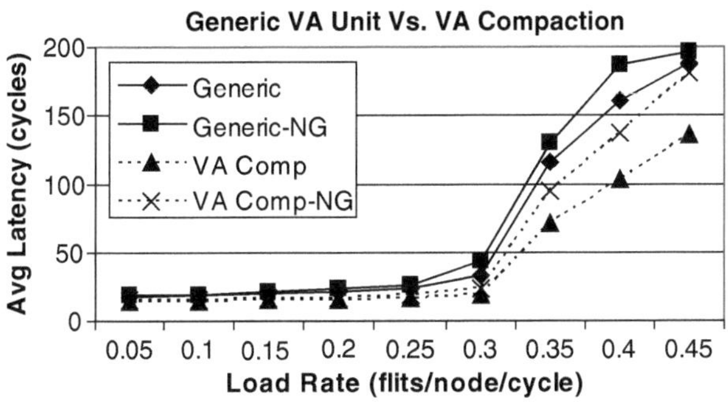

Fig. 6.13 Effect of no-grants on generic and VA Compaction modules

in an NoC router, cutting its critical path in half will enable substantial increases in the overall operating frequency of the router. Furthermore, the notable decrease in the VA's critical path allows the designer to leave enough slack to counter any PV-induced delays in the single-stage arbiters while still being clocked at a much higher frequency than the generic implementation. (2) The number of required arbiters is dramatically reduced. This translates into a 77% decrease in area overhead and power consumption in the VA unit. (3) The resulting configuration is PV-resilient by eliminating the cascade of arbitration logic; without cascaded logic, all inputs to the single-stage arbiters arrive simultaneously, since they now originate from registers. As explained in Section 6.3.2.2, synchronous sources eliminate the possibility of double-grants. Hence, ***VA Compaction provides full immunity from packet losses afflicting generic VA units*** (illustrated in Fig. 6.6). The important result highlighted in Fig. 6.13 is that VA Compaction gains enough in performance to allow it to outperform the *ideal* generic router even in the presence of no-grants ("VA Comp-NG" vs "Generic" curves). In general, *VA Compaction outperforms conventional implementations by about 23% in the presence of PV artifacts* ("VA Comp-NG" vs "Generic-NG" curves).

Hence, ***by re-architecting the entire VA unit, we were able to design a PV-resilient structure which offers significantly better performance while being much more area- and power-efficient than a conventional VA module.***

6.4.3 SA Folding Mechanism

The Switch Allocation (SA) unit coordinates the operation of the crossbar, as described in Chapter 2. Switch allocation is performed in two stages, just like VC Arbitration (VA): a local stage (SA1) and a global stage (SA2), as depicted in **Error! Reference source not found.**(b).

Similar to the VA module, PV-induced delay in the first-stage arbiters will result in glitches at the outputs of the second-stage arbiters. As highlighted in

Section 6.3.2.2, the most serious effect of PV-induced upsets will be double-grants. Upsets in the SA module inevitably lead to significant performance degradation, because the SA process is a "per-flit" operation, i.e., it is needed for all flits traversing the router (as shown in Fig. 6.7 and explained in Section 6.3.2.2).

Unfortunately, a process similar to VA Compaction is not applicable in the SA case. The two arbitration stages are mandatory in efficiently allocating the crossbar resources. A possible safeguard against PV-induced glitches in the SA unit is to synchronize the inputs of the second-stage arbiters by placing registers between the two arbitration stages. This amounts to a splitting of the SA unit into two distinct pipeline stages. While this approach would provide PV resilience, overall router performance would suffer, since there is now an additional pipeline stage.

To overcome this limitation, the author proposes a technique called ***SA Folding***. SA Folding splits the SA process into two phases (local and global), but instead of cascading the two pipeline stages, it parallelizes them and performs both simultaneously using speculation. If the speculative execution of SA2 succeeds, then the SA1 result is discarded. If speculation fails, then the result of SA1 is used non-speculatively in the next clock cycle. The idea of SA Folding is depicted abstractly in Fig. 6.14, with a detailed architectural view shown in Fig. 6.15. Separation of the two stages is facilitated by a *P*-deep FIFO register, as will be explained later on. As illustrated in Fig. 6.15, a newly arriving flit undergoes SA1 and SA2 simultaneously. If both SA1 and SA2 are successful, the result of SA1 is discarded (through the FIFO pointer logic). The FIFO read pointer is not updated until a flit succeeds in SA2. During any given cycle, at most one flit will be performing speculative SA2 per port, since at most one new flit can arrive in one clock cycle in each port. In Fig. 6.15, the darkened portion of the logic represents our SA Folding additions to the generic SA unit.

Note that SA Folding integrates nicely with VA Compaction, because with the VA unit's latency almost halved (see previous section), SA becomes the new bottleneck. Hence, SA Folding comes to the rescue by parallelizing the two arbitration stages, and effectively halving the SA unit's critical path.

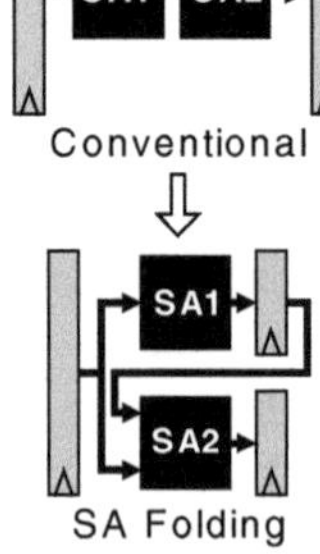

Fig. 6.14 The SA Folding concept

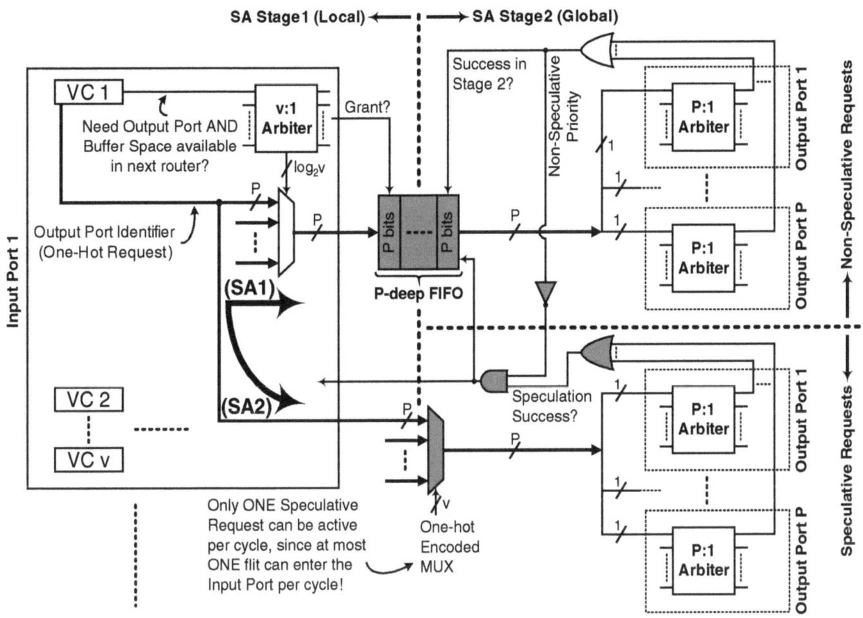

Fig. 6.15 Detailed view of the SA Folding architecture

SA Folding uses two different sets of SA2 arbiters: one for the speculative requests, and one for the non-speculative ones. This is NOT additional overhead, since dual SA arbiters are already present in high-performance conventional routers which parallelize the SA and VA stages [48,49]. SA Folding simply utilizes existing resources to improve the efficiency of the SA stage. Just like in a traditional SA unit, priority is always given to non-speculative requests (see Fig. 6.15).

The effectiveness of speculative schemes is linked to the speculation success rate. In the case of SA Folding, speculation is almost always successful under light traffic load, since the probability of two packets requesting access to the same output port at the same time is very low. At higher traffic loads, speculation often fails. However, this is inconsequential to performance, since at high loads even conventional SA fails often (because of increased contention). Even if SA fails for some flits, it is successful for others, so throughput remains unaffected. This is a very significant characteristic of router architectures. Building on this trait, SA Folding is in fact better than the conventional cascaded approach, because the clock period is now much shorter. Therefore, a wasted cycle in SA Folding is much shorter than a wasted cycle in a conventional router.

Moreover, the proposed SA Folding mechanism solves another inefficiency associated with the traditional two-stage SA process. To improve fairness in arbitration, once a winner is decided in SA1, the arbiter is "locked" until the winning request succeeds in SA2 as well [48]. In this fashion, when an SA1 winner loses in SA2, it does not re-arbitrate in SA1. While this method is fair to a particular flit, it adversely affects throughput, because SA1 cannot operate in parallel with SA2 due to the locking mechanism. This effect is more pronounced under heavy load, when a flit may lose multiple times in SA2 (when many input ports request access to the same output port). To overcome this issue, SA Folding separates the two stages with a *P*-deep FIFO register, as shown in Fig. 6.15. If a flit keeps losing arbitration in the second stage, operation in SA1 is uninterrupted. A flit will lose at most *P* times (since SA2 arbiters are *P*:1); hence, the *P*-deep FIFO register.

While SA Folding increases the size of the SA unit itself by around 35% (due to the *P*-deep FIFO), the effect on the overall router is negligible, because the SA unit represents a very small portion (in terms of area and power) of the entire router. Synthesis results indicate that SA Folding incurs a mere 1% area and power overhead *over the entire router*. Remember, a *P*-deep *P*-bit FIFO corresponds to only 25 bits for $P=5$ (i.e., five input/output ports); by comparison, a *single* flit slot is 128 bits wide. Thus, since the flit buffers dominate the router (80 slots per router, as explained in Section 6.3.1), the SA Folding overhead is insignificant.

SA Folding ensures that all inputs to the SA arbiters arrive from registers. Just like VA Compaction, this trait eliminates double-grants and their associated packet losses. Furthermore, this protection does not come at the expense of performance. On the contrary, the more efficient SA Folding mechanism provides a 10% improvement in performance, as shown in Fig. 6.16a ("Generic" vs "SA Fold" curves). If we take into account that SA Folding decreases the critical path of the router, then clock frequency may increase. Figure 6.16b shows that the performance improvement due to SA Folding increases to around 30%, if the clock frequency increases by 20%. Finally, *in the presence of PV-derived no-grants, SA Folding achieves 11% higher performance assuming equal clock frequency* ("Generic-NG" vs "SA Fold-NG" curves in Fig. 6.16a), *or 29% higher performance with a 20% higher clock frequency* ("Generic-NG" vs "SA Fold-NG" curves in Fig. 6.16b).

Hence, ***by parallelizing the two arbitration stages, the SA Folding mechanism protects the switch allocation process from PV-induced double-grants and flit losses. Moreover, performance actually improves because of (1) a more efficient arbitration scheme, and (2) the halving of the SA critical path delay, which allows for a much higher operating clock frequency.***

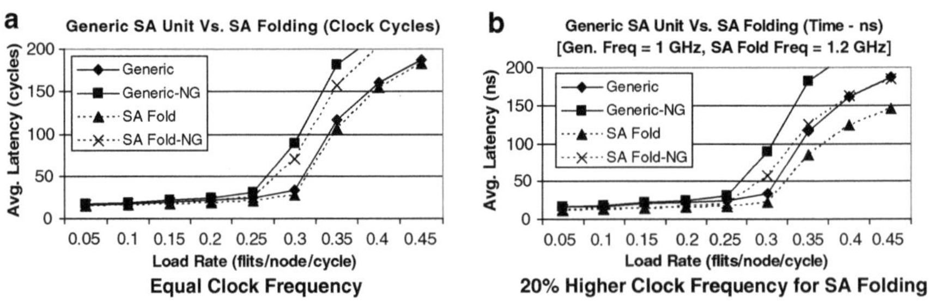

Fig. 6.16 Effect of no-grants on generic and SA Folding modules. (**a**) Equal clock frequency. (**b**) 20% Higher clock frequency for SA Folding

6.5 Chapter Summary

Networks-on-Chip are increasingly becoming susceptible to an emerging vicious threat: Process Variation. Process variations in a chip are a result of manufacturing imperfections, which can lead to substantially degraded performance and even erroneous behavior.

Given the de-facto importance of the interconnect in the overall system it is imperative to protect the network from PV artifacts. In this chapter, the author presented the first comprehensive study of the impact of PV on each of the major components of the NoC. After identifying the possible consequences and their severity, a new router design, ***SturdiSwitch***, was proposed, which includes three novel microarchitectural features to fight PV-induced upsets. Through cycle-accurate simulation and synthesized implementations of the proposed design, the author demonstrated the effectiveness of SturdiSwitch in *providing a PV protective layer over the entire network operation with no loss in performance*. In fact, by re-architecting several key router components, *SturdiSwitch manages to outperform conventional implementations by as much as 23%, while being more area- and power-efficient.*

Part II
MACRO-Architectural Exploration

Chapter 7
The Quest for Scalable On-Chip Interconnection Networks: Bus/NoC Hybridization [15]

Systems-on-Chip (SoCs) have become the design methodology of choice for a multitude of applications, spawning a number of on-chip communication-related challenges. So far, the two dominant architectural choices for implementing efficient communication fabrics for SoCs have been *transaction-based* processor interconnect *buses* and *packet-based Networks-on-Chip (NoC)*. Transaction-based buses have in-built design choices that constrain their application, making them unsuitable as efficient SoC interconnects. Networks-on-Chip, on the other hand, suffer from inconsistent latencies and deterioration of performance at high data injection rates due to their network pedigree.

In this chapter, we aim to address this disconnect between buses and NoCs in an attempt to develop a highly scalable and efficient interconnection framework that captures the intricacies of both heterogeneous Multi-Processor SoCs (MPSoC) and homogeneous Chip Multiprocessors (CMP). Before embarking on this quest, we first evaluate thoroughly existing interconnect types, including the newly-proposed Code-Division Multiple-Access (CDMA) interconnects. Building on this exploratory phase, a new transaction-less bus, the Dynamic Time-Division Multiple-Access (dTDMA) bus, is then developed and analyzed, and its performance quantified and compared to existing buses and NoCs. The dTDMA bus was specifically architected to overcome the weaknesses of transaction-based buses and be amenable to bandwidth-hungry SoC applications.

Finally, a hybrid SoC interconnection topology that employs both the dTDMA bus and an NoC is presented. The hybrid topology provides a scalable, low-latency interconnect by exploiting the speed of the dTDMA bus and the scalability of the NoC. The hybrid interconnect is shown to produce significant improvements in both latency and power consumption over a pure NoC interconnect.

7.1 Introduction and Motivation

The result of the recent onslaught toward complex, multi-core SoCs is that previously distinct design paradigms are being incorporated into single monolithic designs. Chip designers find themselves integrating analog, digital, and memory circuitry – in

C. Nicopoulos et al., *Network-on-Chip Architectures: A Holistic Design Exploration*, Lecture Notes in Electrical Engineering 45,
DOI 10.1007/978-90-481-3031-3_7, © Springer Science+Business Media B.V. 2009

essence, an entire system – onto a single die. All processing elements in an SoC must be connected in some fashion; ideally with a power- and performance-efficient interconnect. The design of such an interconnect is proving to be a crucial part of system design, greatly impacting both power consumption and performance [12]. A poorly conceived interconnect can inhibit performance and/or consume significant amounts of power; both are problems in high-performance and embedded designs.

Several on-chip interconnection schemes have been conceived, including crossbars, buses, and networks; however, to date, two of these have dominated the design space – processor interconnect buses and networks [12]. Processor interconnect buses could be considered the more traditional of the two, having been in use for decades. By sharing one set of wiring resources in a controlled manner, a bus permits multiple participants to send and receive data from each other. Coordinating access to the bus is an arbiter, to which bus participants send requests for bus access. Existing industry-standard buses include designs from ARM (AMBA [120] and AXI [121]), IBM (CoreConnect [122]), and STMicroelectronics (STBus [123]).

The general design principles and behavior of buses is well known, and as such, a diminishing amount of research is aimed at them. However, there are a number of improvements that can be made to existing bus designs that can better suit them for use in SoCs. In this chapter, we propose a new transaction-less Time-Division Multiple-Access (TDMA) bus with dynamic timeslot allocation – the dTDMA bus – in order to circumvent many of the problems associated with traditional bus architectures. In dTDMA, a bus arbiter dynamically grows and shrinks the number of timeslots to match the number of active transmitters. The dynamic nature of this architecture guarantees non-blocking operation, which substantially alleviates the contention problem inherent in bus architectures. This, coupled with its single-hop communication nature and lack of transaction-style arbitration, allows for low and predictable latencies. Hence, dTDMA provides a two-fold advantage over current SoC interconnects: it tackles the contention and transactional issues of existing bus architectures, and avoids the multi-hop overhead of NoC solutions. We show that these traits allow the dTDMA architecture to outperform traditional bus architectures, as illustrated by an example comparison with the AMBA bus.

The other dominant on-chip interconnect architecture in modern SoC designs is the NoC, as already described in the preceding chapters of this volume. Both buses and networks have advantages over each other, but neither one alone can solve the interconnect demands of modern SoCs. While buses tout low latency due to the nature of their single-hop communication, they do not scale well, meaning that a bus cannot be used to connect a large number of processing elements. On-chip networks (especially mesh networks), on the other hand, have the ability to connect a larger number of processing elements with a single interconnect but can introduce high and variable latencies due to the multi-hop nature of network communication and the dynamic behavior of certain routing algorithms.

Though networks and buses may, at first, appear to be mutually-exclusive interconnect solutions, it is possible to use a combination of the two to produce a single, efficient interconnect. In this chapter, we propose a new hybrid interconnect that exploits the benefits of dTDMA for small configurations and the scalability of

NoCs. In our hybrid interconnect, processing elements exhibiting communication affinity are grouped together on dTDMA buses, while the remaining elements communicate through an NoC. We show that this hybrid dTDMA/NoC system significantly outperforms pure NoCs.

7.2 Exploration of Existing On-Chip Bus Architectures

In this section, traditional System-on-Chip interconnects, their design, operation, and constraints are discussed. An original, thorough analysis of the feasibility of CDMA interconnects is presented here. This part serves as a motivation for the development of the dynamic TDMA bus, which is discussed in detail in the following section.

7.2.1 Traditional Bus Architectures

Shared buses like AMBA and CoreConnect have long been the traditional interconnect of choice for SoC designers because of their simple topology and extensibility [124]. By providing a standardized interconnect, these shared buses allow designers to stitch together various hard and soft IP cores in an SoC. The general structure of such buses consists of some number of masters and slaves, an arbiter, the data bus, and the address bus. In the context of SoCs, masters are Processing Elements (PEs) that are allowed to initiate communication, and slaves are the PEs that can only respond to masters. The bus arbiter coordinates all communications between masters and slaves through a series of control signals.

The data bus itself is not a single monolithic broadcast bus, but instead two buses – read and write – that are each separated into two parts by multiplexers and OR gates in AMBA and CoreConnect, respectively. The address bus is similarly split into two portions by multiplexers and OR gates.

Both AMBA and CoreConnect are *address-mapped* buses, meaning that slaves are bound to certain portions of the memory space. Such an addressing scheme is very suitable for processor/memory configurations. When a processor (master) wishes to perform a memory operation (with some slave), it simply asserts the address that it wants to read from or write to. A simple address decoding allows the appropriate slave to respond to the request.

AMBA and CoreConnect are also *transaction-based* buses, that is, communication over the bus occurs in atomic transactions. Bus transactions occur in two phases – the address phase and the data phase.

When a slave is unable to respond to a request quickly (e.g., a slow memory performing a read), the slave can request a *transaction split*. The split breaks atomicity, allowing the transaction to be completed later while releasing the bus to other masters in the meantime. However, when the slave wishes to complete the request, it must undergo an arbitration delay to gain access to the bus.

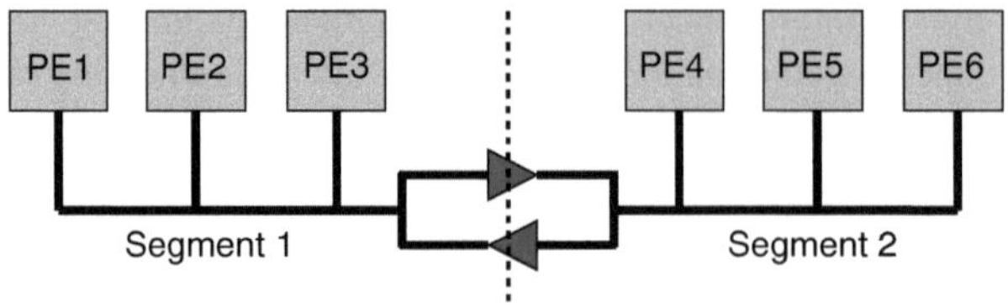

Fig. 7.1 Bus segmentation

7.2.1.1 Bus Segmentation

A problem present in traditional shared bus architectures is that a single bus transaction consumes all of the bus resources. An example of this is as follows: Several PEs (1 through 6) sit on a linear bus. If PE1 is communicating with PE3, then no other communication can take place on the bus. A common solution to this – known as segmentation – is to break up the bus into segments that can independently support transactions [125]. If the bus in the previous example were segmented into two halves, then PE1 and PE3 could communicate on one segment while PE4 and PE5 communicate on the other segment. Figure 7.1 shows how segmentation would be applied to this example.

The SAMBA-BUS [124] pushes bus segmentation to the limit by providing a segmentation point at each PE on the address and data buses. An additional improvement that SAMBA-BUS makes over traditional shared bus architectures is that even if a PE does not win the arbitration, it may still be allowed to access the bus, based on a set of rules which prevent overlapping communication from occurring. The downside to SAMBA-BUS's improved architecture is that the bus suffers from increased delay due to multiplexers in the critical path, and increased interface logic at each PE.

7.2.1.2 Bus Arbitration

Arbitration policy plays a large role in determining the performance of buses, and as such, LOTTERYBUS [126] is an improvement to arbitration policy. LOTTERYBUS performs arbitration by randomized selection of a bus master based on the master's ownership of a "ticket." Each master can be statically or dynamically assigned a number of tickets. Upon arbitration, the arbiter chooses a ticket in a uniform random fashion from a pool of the tickets of masters with pending requests – a process that reduces the starvation of low-priority PEs. LOTTERYBUS can also effectively guarantee bandwidth to certain PEs by allocating more tickets to them.

However, the drawback to the LOTTERYBUS is a relatively high hardware requirement. The arbitration that LOTTERYBUS performs is simple, yet the hardware implementation is quite complicated. A simpler method of reducing low-priority PE starvation is presented in Section 7.3.

7.2.2 TDMA Buses and Hybrid Interconnects

Though most traditional shared bus architectures can implement some form of time-division multiple-access arbitration inside a more complex arbitration protocol (as a means of contention resolution), there have been some recent bus architectures that use TDMA as the sole form of contention avoidance and contention resolution. Instead of implementing TDMA inside of another protocol (e.g., using TDMA to resolve multiple access requests in an AMBA bus), pure-TDMA buses grant bus ownership in a periodic manner without making use of any other ownership rules or protocols.

A TDMA bus, at its most basic level, functions by allowing processing elements to transmit on the shared bus wires in a round-robin fashion. For example, in a bus consisting of *i* PEs, PE_i is allowed to transmit on the bus every *i* cycles. Access to the bus is divided through time, as opposed to being divided through codewords, in the case of CDMA which will be examined later on. The allocation unit in a TDMA system is the *timeslice*, or a portion of time, a set of which constitutes the *superperiod.*

Traditional TDMA systems use a fixed number of timeslices (i.e., fixed superperiod), regardless of the number that are allocated. This method has the advantage of being simple to implement (each of *i* transmitters simply transmits once every *i* cycles) but wastes bandwidth when some timeslices are unallocated. A fixed-superperiod TDMA system would perform poorly under bus traffic with a time-varying profile. Specifically, there will be significant portions of time where some number of PEs do not wish to transmit, and, consequently, their timeslices exist unallocated, squandering bandwidth, much to the chagrin of other PEs that wish to transmit data.

To overcome bus underutilization, unused timeslices can be re-assigned to PEs that are ready to transmit [127]. This has a positive effect on utilization but often results in unfair timeslice assignments. Figure 7.2 shows how in a system with eight timeslices, there is no fair way to assign the unassigned timeslices to the active transmitters.

The SiliconBackplane III TDMA-based SoC interconnect from Sonics, Inc. [128] implements timeslice re-assignment through a two-level timing wheel. Each PE on the bus is assigned one slot on the timing wheel. When that timeslice is the current slice and the PE has a pending transfer, bus ownership is granted. If the PE

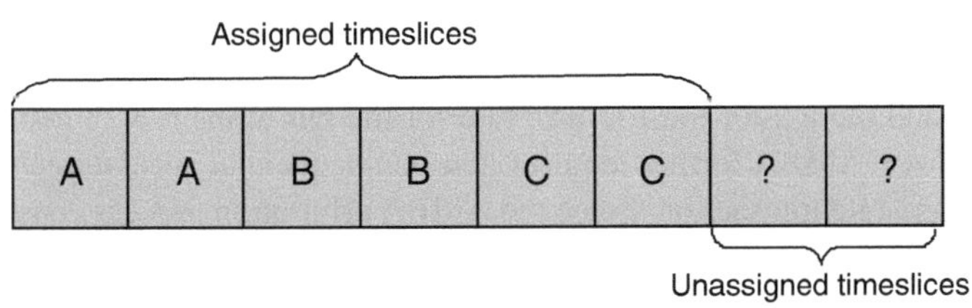

Fig. 7.2 Unfairness caused by timeslice reassignment

does not have a pending transfer, then a second level of arbitration assigns the timeslice to some other PE with a pending transfer. In this way, SiliconBackplane III reclaims unused timeslices, reaching a maximum bandwidth utilization of 90%. While such a maximum utilization may seem high, a more efficient TDMA bus is developed in Section 7.3.

SonicsMX [129] uses a hybrid shared-bus/crossbar approach to improve performance and reduce power consumption. However, it employs a transactional model of communication, which, as we will demonstrate later, introduces inefficiencies.

Niemann et al. [130] have proposed a bus/network hybrid specifically for the application of network processors. However, little insight into design decisions and no comparison with existing technologies are offered.

7.2.3 Constraints of Traditional Buses

The preceding analyses of existing SoC interconnect standards reveals several important common constraints on the use of these interconnects. Firstly, AMBA, CoreConnect, products from Sonics, Inc., and even the enhanced buses, SAMBA-BUS and LOTTERYBUS, all rely on the master–slave model of processing and communication. This model, defines masters as a set of processing elements that can initiate communication and slaves as the remaining set of processing elements, which cannot initiate communication. Within the context of a shared-memory multiprocessor system, such a model functions well – processors are masters that initiate communication (memory accesses) with some set of slaves (memory banks). However, in SoCs, where communication could easily be bi-directional, the master–slave model imposes unacceptable restrictions on system flexibility. One solution could be to make each PE a master and a slave, however, many interconnects place an upper limit on the number of masters that can exist (16 in CoreConnect and AMBA) [122].

A second restriction imposed by existing SoC interconnect standards is that the bus necessarily be memory-mapped. Again, while this model may work well for strictly processor-memory interaction, it introduces inefficiencies that are undesirable in more heterogeneous SoC designs. Memory-mapping alone is not cause for inefficiency in communication, as it is merely a form of addressing destinations. However, AMBA – for example – imposes arbitrary address boundaries on transactions. For example, AMBA limits sequential transfers (accessing memory locations incrementally) to a 1 kB address range (CoreConnect does not impose such boundaries) [120]. When a transfer crosses an address boundary, the transaction must be terminated and the master must re-arbitrate for the bus at the risk of losing ownership. Moreover, AMBA further requires that non-sequential accesses each be arbitrated as a separate transaction. Since the AMBA arbitration process consumes four cycles from request to address transmission, and five cycles from request to data transmission, excessive arbitration overhead may result from long or non-sequential memory accesses.

A third restriction stems from the fact that communication between processing elements occurs in transactions, which consist of two phases – address and data. For data traffic that does not require an address (e.g., processor-to-processor data streaming), the address cycle increases the request-to-data latency and essentially wastes bandwidth.

In summary, traditional shared-bus architectures suffer, when applied to the SoC design space, from their reliance upon the master–slave, memory-mapped, transactional model which is inadequate for the heterogeneous nature of SoC communication. A new interconnect model is necessary to support efficient communication between and straightforward integration of the various components in modern SoC designs.

7.2.4 CDMA Interconnects

Though most commonly associated with wireless communication, Code-Division Multiple-Access, or CDMA, has recently been introduced as an alternative to traditional bus architectures. CDMA differs from all other modulation schemes in that it permits all transmitters to transmit at once by using a unique coding sequence for each. The intended advantage to using CDMA as an interconnect is twofold: (1) support for dynamically-reconfigurable interconnects, and (2) various benefits of simultaneous communication [131,132]. The foundation upon which CDMA is built consists of what are known as spreading sequences, spreading codes, or codewords.

Consider the following scenario: Two transmitters, Tx_A and Tx_B, can modulate two data streams, D_A and D_B, using two spreading sequences S_1 and S_2. The two modulated data streams are allowed to interfere on one communication channel (the modulated data streams sum onto the channel). Two receivers, Rx_A and Rx_B, listen to and demodulate the communication channel using the same two spreading sequences, S_1 and S_2. Sequences S_1 and S_2 are chosen to have minimal cross-correlation and maximal autocorrelation. Thus, data modulated with one sequence will not destructively interfere with the data modulated by the other sequence, and may only be demodulated using the sequence that modulated it.

During modulation, each data bit is spread by multiplying it by each bit of the spreading sequence, producing what are known as chips [133]. Modulation and demodulation for a single data bit are performed over the period of the codeword, T_s. Thus, assuming an N-bit spreading sequence, if data bits occur at some frequency f_{data}, then the codeword bits (chips) must be multiplied by each data bit at the frequency $f_{chip}=N_{chip}*f_{data}$, to maintain data throughput. In other words, if the frequency of the communication channel can be increased by a factor equal to the length of the codeword, then the original data throughput is maintained. However, considering the capacitive load presented by the CDMA bus, increasing the bus frequency by a factor of N_{chip} may be infeasible. The bus could be pipelined, but there would still be a significant dynamic power draw from the bus lines switching at such a high frequency, which is very undesirable for SoCs, many of which are embedded systems.

A large portion of the problems associated with CDMA stem from the choice of codewords/sequences, including issues related to type, size, and generation. Spread spectrum systems like CDMA generally make use of two types of spreading sequences – orthogonal codes and PN sequences [134,135].

Walsh codes are the easiest type of orthogonal codes to generate. However, the lengths of Walsh Codes can only be powers of two, resulting in many cases where data is being over-spread. Figure 7.3 shows how throughput varies with the number of active transmitters, *P*, when using Walsh codes. In a periodic manner, a significant portion (up to 50%) of the bandwidth is wasted. The loss in throughput is the result of overspreading with a larger-than-necessary codeword – using a codeword of length 2 $[log_2 P]$, where one of length *P* is optimal.

This inefficiency of Walsh codes in utilizing available bandwidth can be mitigated through the use of the aforementioned second type of spreading sequences, called PN sequences. Gold codes and Kasami sequences are two examples frequently encountered in spread-spectrum systems [136]. The lengths of these codewords are not restricted to powers of two; the codewords can be of any length, thus improving bandwidth efficiency over Walsh codes. However, both Gold codes and Kasami sequences are produced from decimated versions of so-called "maximal sequences," which are longer codewords. Consequently, generation hardware must be clocked at a multiple of the chip frequency to ensure on-the-fly data modulation.

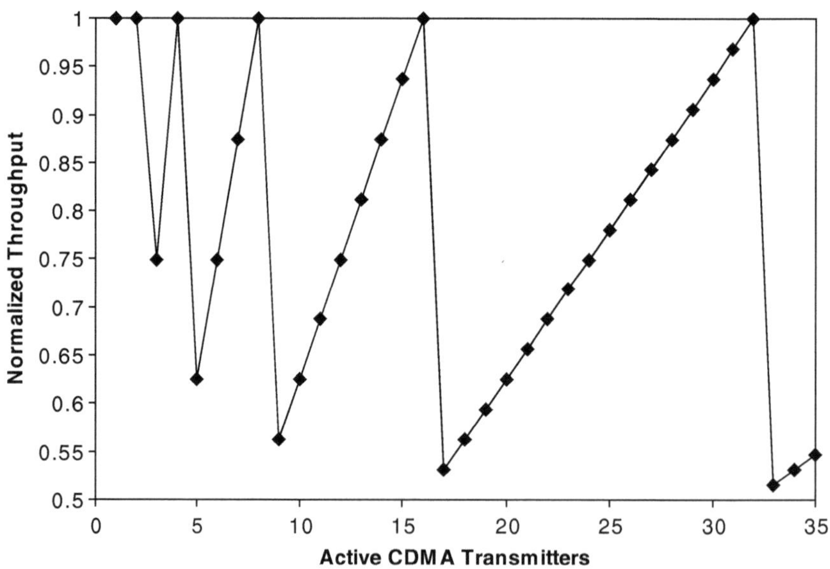

Fig. 7.3 Normalized CDMA throughput using Walsh orthogonal codes

The final hurdle in implementing CDMA as an on-chip interconnect is how to represent the communication channel, S_{tot}. Traditionally, and as a result of its wireless heritage, CDMA uses an analog communication channel. Transmitters simply broadcast their modulated signals which interfere to form S_{tot}. In wired systems, the channel may also be represented analogously [137–139]; however, this implies extensive use of mixed-signal (analog and digital) VLSI circuitry, which severely complicates the design, implementation and verification of the overall system. The alternative is to use binary (digital) representation of S_{tot} [131,140]. However, to represent all possible values that S_{tot} may take on, multiple bits are needed. Specifically, to represent a signal that can take on n values, $b = [log_2\ n]$ bits are necessary. This wire multiplicity turns into significant overhead in wide on-chip buses, where *each* data bit would require b wires.

Hence, CDMA has a number of drawbacks that detract from its intended benefits – simultaneous communication and dynamic reconfiguration. Data spreading reduces throughput, codeword generation is non-trivial, and physical restrictions such as wiring density and mixed-signal integration further limit the appeal of using CDMA as an on-chip interconnect.

7.3 The Dynamic Time-Division Multiple-Access (dTDMA) Bus

Having established that standard bus designs impose several constraints upon their application due to their inherent master–slave, memory-mapped, and transactional mode of operation, we thereby contend that the use of these standard buses as interconnects for heterogeneous SoC designs is inefficient. An example SoC highlighting these inefficiencies is a modern cable or satellite television Set-Top Box (STB) (or an embedded multimedia processor such as Nomadik [141]), which performs simultaneous decoding of audio, video, and data content (electronic program guides, interactive programming, etc.). The key difference between communication in a system like this and the types of systems for which the standard shared buses were developed is that in the case of the STB, communication occurs in long streams, possibly involving multicasting, instead of intermittent bursts. Transaction-based buses fail in regards to data streaming because all communication must be divided into transactions, each undergoing an arbitration and its associated delay. Also in regards to data streaming, the use of a memory-mapped bus architecture imposes inefficiencies (arbitrary address boundaries on transactions and two-phase transactions) which could be avoided if a more appropriate addressing scheme were used. Another aspect of the STB communication pattern where standard shared buses fall short is bi-directional communication. Recall that AMBA, CoreConnect, and others use the master–slave model of communication, which effectively limits the initiation of communication to masters. In the STB and other SoC designs, there are instances where bi-directional communication is necessary (e.g., a video codec core that outputs when decoding and inputs when encoding).

Furthermore, existing bus architectures employ relatively complex protocols that make designing a compliant interface a non-trivial task, requiring compliance checking and certification [142]. dTDMA was specifically architected with a very simple protocol to facilitate easy integration in multi-core SoCs.

The following section details the design of the dTDMA bus and demonstrates how it overcomes the limitations of existing SoC interconnects in a new and elegant manner.

7.3.1 Operation of the Dynamic Timeslot Allocation

The basic operation of the dTDMA bus is as follows. When a PE needs to transmit, it asserts its active signal to the arbiter to request a timeslot. The arbiter then decides (using any number of techniques) on a timeslot assignment for each PE and produces a new configuration for each active transmitter and receiver before the next clock cycle. On the next clock edge, active transmitters and receivers load the new timeslot configuration data and continue operation. When a transmitter is finished communicating, it de-asserts its active signal, and the arbiter de-allocates a timeslot in the same manner that it allocates them. Figure 7.4 illustrates this operation based on transmission requests.

This method of dynamic allocation always produces the most efficient timeslot allocation – timeslots are never wasted. That is, a timeslot will only exist if it is allocated to a PE, and a timeslot will never be allocated unless the PE is ready to transmit and requests a timeslot. Because of this, the dTDMA bus is nearly 100% bandwidth efficient (99.9%), thus making the bandwidth utilization equal to the injection rate. The small overhead is due to the initial communication delay of one cycle when a timeslot is allocated.

Dynamic management of timeslices frees the system designer from having to manually optimize the assignment of timeslices as would be done in a fixed-superperiod system, similar to the process appearing in [143]. Such a task could become quite complex considering the variation in on-chip traffic, especially for multimedia processing on MPSoCs [144].

Different methods can be used to assign timeslots to transmitters. The method shown in Fig. 7.4 allocates the first timeslot(s) to the newly-transmitting PE(s).

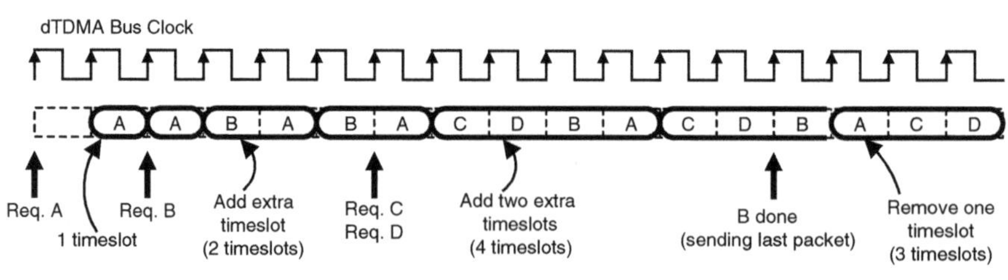

Fig. 7.4 Example of dynamic timeslot allocation

Other methods can schedule timeslots based on the status of transmit buffers or on the length of wait time, as will be described later on. Additionally, various Quality-of-Service (QoS) techniques can be applied to the timeslot assignment algorithm. By assigning a certain percentage of timeslots to a PE, a percentage of bandwidth can be guaranteed. Priority schemes and absolute bandwidth guarantees are also possible by blocking timeslot allocations.

7.3.2 Implementation of the dTDMA Bus

The general architecture of the dTDMA bus is a single broadcast bus connecting all PEs, which have simple transceiver units whose operation is coordinated by an arbiter through a series of control signals. The bus interface (Fig. 7.5) of each PE consists of a transmitter and a receiver connected to the bus through a tri-state driver. The tri-state drivers on each receiver and transmitter are controlled by independently programmed fully-tapped feedback shift registers. Multiplexers between each stage allow for parallel load and a variable length shift register – an n-stage shift register can be programmed to any length from 1 to n.

Destinations are represented as a multiple-hot encoding, requiring n lines (one for each PE). Similarly, the timeslot configuration data requires n lines. The control signals require *log(n)* lines because they are the select lines for the decoder that sets the feedback path (Fig. 7.5).

During operation, the shift register rotates one stage each clock cycle. The head of the register acts as the enable signal to the tri-state driver. Accordingly, if there are five allocated timeslots and a certain source PE requests bus access and is assigned timeslot 2, the source PE and all destination PEs would receive the initialization data {00100}. Two clock cycles later, the PEs shift registers would contain {10000}, enabling the tri-state drivers, driving the bus, and sampling the bus. On the next rising edge of the clock, the destination PE registers the data before the bus data and shift registers change. The shift registers load new data only when the timeslot allocation changes (i.e., when a PE initiates or ends transmission) and simply operate in rotate-mode otherwise.

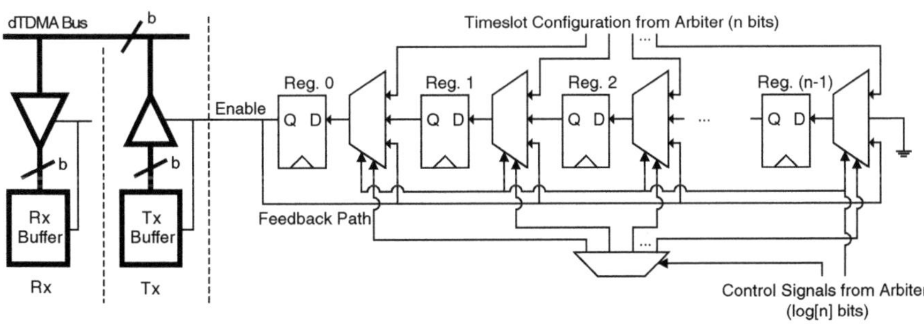

Fig. 7.5 Architecture of the dTDMA receiver and transmitter

Because of its broadcast design, the dTDMA bus can easily support multicasting. Multicasting has applications in Multi-Processor SoCs (MPSoC), such as cache coherency protocols [145] and Single Instruction Multiple Data (SIMD) processing. Support for multicasting is simple and inherent in the arbitration protocol; introducing no additional latency.

To support many-to-one communication, first, the transmitting PEs must always prepend data with a start flag which is unique to them. This start flag is always paired with an end flag to signal the end of the transmission. Second, receivers must listen to multiple timeslots; support for this is built into the arbitration protocol. Third, receivers must be able to perform simple message reassembly as they may receive more than one message at a time. This reassembly process is no more involved than what is already required with an NoC interconnect or the AXI bus [121].

Using the Berkeley Predictive Technology Model in a 70 nm process [146] and scaled RAW PE tile [20] of square dimension 1.867 mm, the maximum frequency of a 9 PE dTDMA bus was calculated to be about 916 MHz. Buses with 4, 16, and 25 PEs can be clocked at 2.846 GHz, 470 and 289 MHz, respectively. The reduction of maximum operating frequency with larger buses is due to increasing RC delay.

7.3.3 Comparison with a Traditional Bus Architecture

There are several different prevalent interconnect structures in MPSoCs, including buses (AMBA, AXI, CoreConnect, STBus) and NoCs (AEthereal, STNoC, Crossbow, and Faust) [147]. In this section, we will show specific instances in which these transaction-based buses will not perform well, using AMBA as an example.

As previously mentioned, AMBA imposes certain restrictions on the nature of addressing behavior, because of its memory-oriented design. In order to operate properly with memories, AMBA transactions occur in two phases – address and data. For processor-to-processor transfers, this two phase transaction is inefficient, since it wastes a cycle sending unnecessary address data.

Instead, the dTDMA bus is transaction-less and address-mapped. By assigning each PE on the bus a unique identifier (0, 1, 2, etc.), PEs can send data to each other without the restrictions of memory-oriented buses. Additionally, multicasting is made trivially simple with an address-mapped scheme.

An example drawback of AMBA as compared to the address-mapped dTDMA bus is that the dTDMA bus only requires a re-arbitration when the destination of a data stream changes. So if PE0 (a processor) is reading data from M1 (a memory), and requests memory addresses that are not sequential or that cross a 1 kB boundary, no re-arbitration is necessary since PE0 is still communicating with M1. AMBA would require a re-arbitration for either of these situations (even running the risk of losing bus ownership). Another advantage of dTDMA is that, in this example, even though PE0 has to re-arbitrate to change destination (e.g., to M2), it will not lose its previously allocated timeslot position, as all the arbiter has to do is

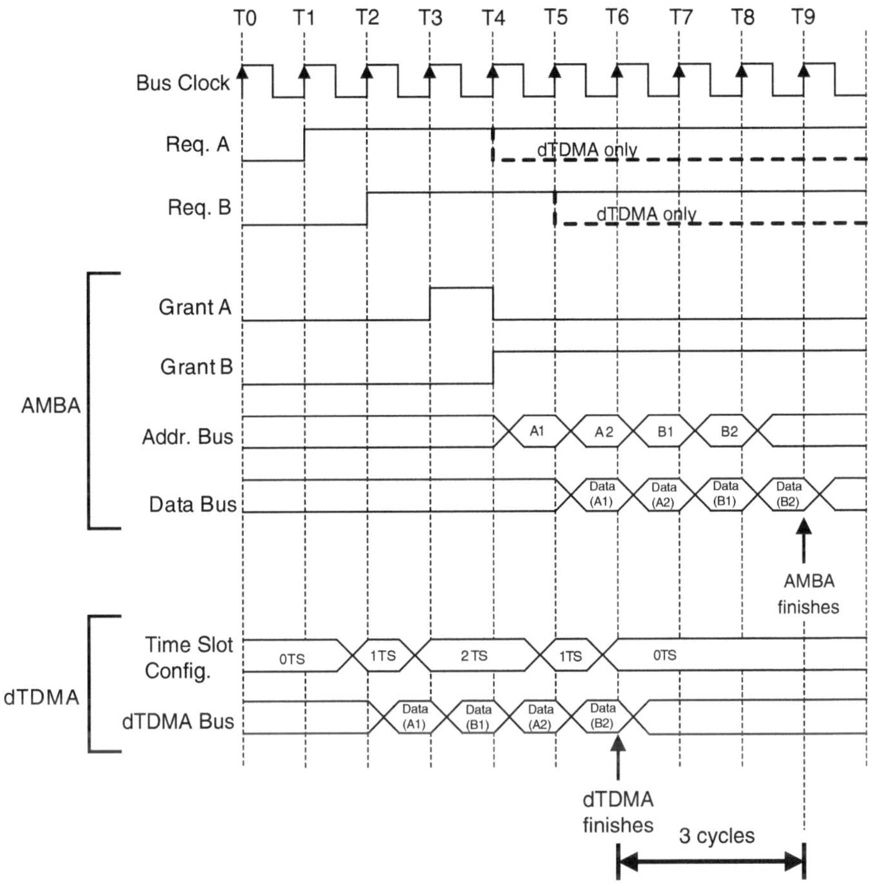

Fig. 7.6 A simple 2-PE transfer on dTDMA and AMBA

reprogram the receivers (M1 and M2). Hence, there is no effective arbitration delay even though a change of destination has occurred.

A simple, illustrative, comparison between AMBA and dTDMA appears in Fig. 7.6. In this scenario, master A requests access to the bus on clock T1, followed by master B requesting access on clock T2. Master A does not receive a grant from the AMBA arbiter until clock T3, while the dTDMA arbiter issues a new timeslot configuration before the end of clock T1. Data transmission commences with clock T5 in the case of AMBA, since T4 is dedicated to the address phase of the transaction. dTDMA allows data transmission to commence earlier, at T2. Master B waits five clock cycles from request to data transfer in the AMBA bus, yet only one cycle in the dTDMA bus. In this scenario, the dTDMA bus completes the transmission of two words from each master three cycles before the AMBA bus does.

dTDMA has the attractive quality of predictable latencies because a PE is guaranteed to wait no longer than the number of active transmitters. In contrast, the AMBA bus may make masters wait an undetermined amount of time before they are granted ownership of the bus. Because of its predictable latencies, the dTDMA bus is also non-blocking, that is, a PE will never be denied access to the bus on account of another PE monopolizing resources. Both of these qualities are important when designing to meet the real-time constraints of multimedia processing applications.

By allocating bus resources as streams instead of transactions, arbitration overhead is reduced for long transmissions. Moreover, the dTDMA arbitration-to-transmission delay is only one cycle, compared to four (for data) in AMBA. The overall effect of moving to a transaction-less, address-mapped bus for streaming data is that fewer arbitrations take place and each can be done more quickly, and a cycle is not wasted in sending address data when it is not needed.

7.3.4 dTDMA Bus Performance

A cycle-accurate simulator was developed in C++ to explore the performance of the dTDMA bus in different situations. Injections are modeled per-PE in a stochastic fashion, varying the mean and standard deviation of the inter-arrival times of messages injected onto the bus. A configuration file is used to describe the operational parameters of each PE. Specifiable parameters include data injection mode (file input, uniform, multimedia, master, slave), injection rate, destination mode (file input, uniform, static, multicast, master, slave), and static destination lists. Two types of traffic injection were used – uniform and multimedia injection. In uniform injection, data arrives with a constant inter-arrival time, implying constant data rate. Multimedia traffic injects data with varying inter-arrival times based on real MPEG-2 traces.

This section outlines the performance of the dTDMA bus with respect to four metrics: throughput, average latency, arbitration policy, and power consumption.

7.3.4.1 Throughput

Throughput is defined as the amount of data delivered to the destination nodes of a network, and is commonly measured in bits per second (bps) [82]. A 512-bit dTDMA bus, operating at 500 MHz, has an ideal throughput of 256 Gbps. This ideal throughput assumes that 512 data bits can be sent on every clock cycle. To measure throughput of the bus, several different bus sizes (4, 8, 12, and 16 PEs) and two different traffic types (uniform and multimedia) were used. For each configuration, the injection rate was varied from 0.1 (10%) to 1.0 (100%) to simulate a full range of use scenarios.

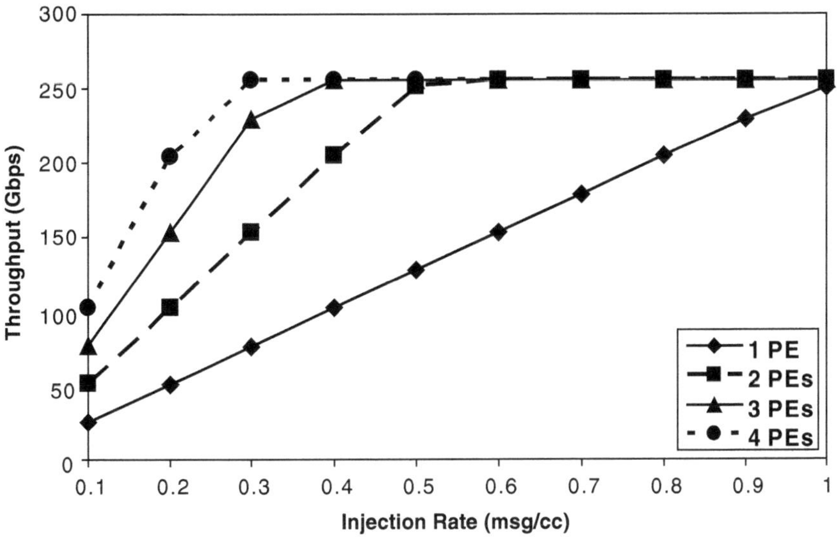

Fig. 7.7 dTDMA bus throughput (uniform injection)

Figure 7.7 shows the throughput of the bus over the injection rate range 0.1–1.0. Trivially, for a single-client bus (1 PE), throughput is equal to the offered load (the injection rate), giving rise to a linear relationship. As more PEs are added to the bus, throughput rises faster with injection rate because, at a given injection rate, more PEs are pushing data through the bus. Beyond an injection rate of about 0.3, the bus becomes saturated with data traffic and hits its ideal (maximum) throughput (256 Gbps).

An important insight can be gleaned from the throughput curve – the dTDMA bus is very bandwidth-efficient. The saturation throughput of the bus is very near (99.99%) the ideal throughput due to the dynamic nature of the timeslice allocation occurring in the bus, which ensures that no cycles are wasted, keeping data flowing on the bus at all times that data is available. Of course, the ideal throughput cannot be reached because there is an access delay incurred by newly-initiated communications. However, this delay happens infrequently (because of the transaction-less design) and is so small (only one cycle, compared to several with traditional buses), that it has a negligible performance impact.

7.3.4.2 Average Latency

Latency – the time required to send data through an interconnect, measured from the injection of the first bit to the ejection of the last bit – is the second metric used to quantify the performance of the dTDMA bus. Since latency is a form of delay measurement, low latency is a desirable trait. However, high latency does not necessarily preclude acceptable performance. For example, in a streaming video

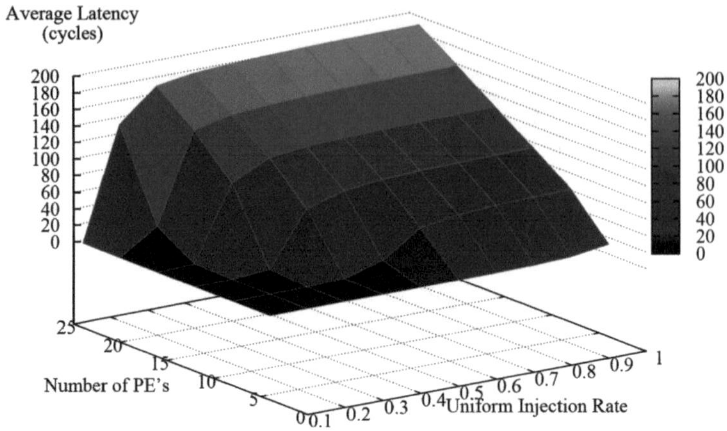

Fig. 7.8 Unicast latencies for several bus sizes

application, the initial delay from first bit in to the first video frame out (latency) may be relatively high, but if the throughput is high enough, then real-time constraints may be met.

Simulations were conducted to quantify the effects of bus size (in terms of bus clients), injection rate and injection type. Without loss of generality, a bus width of 512 bits was assumed here. A *message* size equal to the width of the bus was used (i.e., 512 bits). The results are presented here.

Figure 7.8 shows the latencies for several bus sizes over the entire range of uniform injection rates using unicast (each source transmitting to one destination) messaging. Latency as a function of injection rate has a typical response characteristic – at low injection rates, the latency is near that of the zero-load latency (the latency in the presence of no contention [82]), and at high injection rates, the latency tends towards infinity. However, the response shown in Fig. 7.8 tends towards infinity for a short period of time, but then approaches an asymptotic maximum latency after the injection rate passes the saturation throughput,λ_S. This portion of the response curve is the result of using a finite input buffer size of eight messages for the bus interfaces. When buffer size is unlimited, messages may sit in the buffer for an infinite period, pushing the average latency towards infinity. However, with a finite buffer size, messages generated by the processing element will eventually fill the buffer to capacity, at which point subsequent messages are dropped until buffer space becomes available. These dropped messages are not included in the calculation of average latency. In reality, messages may *not* be dropped, because a full buffer will cause the transmitting processing element to block until the queue empties.

The maximum latency for a 24-PE bus (Fig. 7.8) is about 190 cycles (189.71 to be exact) and is verifiable through the following calculation. In an *N*-PE *saturated* bus, each PE transmits at a rate of 1/N, and on each transmission it moves the last message in its full transmit buffer one position closer to transmission.

Thus, a message starting from the back of an *n*-deep transmit buffer at time t_0 (this is the message that will exhibit maximum latency) waits *n/(1/N)* cycles before being transmitted. When each PE on the bus maintains a full transmit buffer due to a high injection rate, each message inserted into those buffers must wait this maximum amount of time before being transmitted. For the case of the 24-PE bus with 8-deep transmit buffers, the maximum latency is 8/(1/24) = 192 cycles; very near the simulation result of about 190 cycles.

As stated in Section 7.3.2, the dTDMA bus supports efficient multicasting. Considering the strong need for multicasting support by a modern SoC interconnect, the dTDMA bus was enhanced with support for it. The addition of multicast support is actually quite simple; greatly simplified by the broadcast nature of the bus. To be able to send to more than one receiver, a transmitter must be able to specify all destinations to the arbiter. As mentioned in Section 7.3.2, each PE has a set of address lines of width equal to the number of PEs on the bus. This allows for multi-hot encoding of destinations. For example, if a PE wished to transmit to PEs 1, 3, and 4 on a 6-PE bus, it would assign its address lines the value {101100}. The arbiter then programs each of the receivers such that they are "listening" to the bus during the source PE's timeslice. The arbiter is capable of programming, in parallel, every receiver to listen to every timeslice through a set of configuration lines, again, of width equal to the number of PEs.

No additional arbitration delay is necessary to support multicasting, and more importantly, no additional data traffic on the bus is necessary either. For these two reasons, the latency response of the dTDMA bus when multicasting is *exactly* the same as when it is unicasting, *regardless of the amount of multicast traffic.* Simulation results (not shown here for brevity) validate this assertion: multicast latencies are identical to unicast latencies when using the dTDMA bus.

7.3.4.3 Arbitration Policy

Arbitration in the dTDMA bus refers to the task of assigning timeslices to active transmitters. As mentioned in Section 7.3.1, there are several ways to perform this assignment, but it is of great benefit to use an arbitration method that reduces starvation.

Three arbitration schemes are studied within the context of the dTDMA bus – Sliding Timeslice (ST), Fullest Buffer First (FBF), and Longest Wait First (LWF). Each method has a unique way of assigning the high-priority timeslice and subsequent remaining timeslices.

The first arbitration method developed was Sliding Timeslice. It was intended as a simple first attempt at starvation *avoidance* through the prevention of any one PE from monopolizing data transmission. As the name suggests, PEs get to transmit based on a "sliding" window which moves through all requesting clients. The first-occurring timeslice (i.e., the first PE to transmit after a re-arbitration) is assigned to the previous second-transmitting PE. For example, if PE4 transmitted first and a re-arbitration takes place, the first-occurring timeslice would be assigned to the next transmitting PE of PE5, PE6, PE7, and so on (the window eventually wraps around to PE0 and continues).

Fullest Buffer First arbitration attempts to provide starvation *prevention* by giving top priority to the transmitter with the fullest buffer. The second-highest priority is given to the second-fullest buffer, and so on. Thus, FBF arbitration attempts to avoid starvation by giving full buffers transmit priority over less-full buffers. However, FBF suffers from a fundamental flaw. If PEs initiate and terminate communication often enough to cause a high frequency of re-arbitrations, those buffers that are not near-full will have to wait a long time to transmit; possibly not even getting the chance to transmit before another communication change causes a re-arbitration. This process will continue until those near-empty buffers become near-full and finally get to transmit. In short, FBF arbitration performs poorly in situations where communication patterns change frequently.

Longest Wait First gives priority to the PE that has been waiting the longest to transmit. This is essentially round-robin with the additional rule that newly-transmitting PEs are placed at the end of the waiting list. Because transmit priority is preserved when transmitting PEs are added or removed, LWF arbitration guarantees that a PE will never wait to transmit for more clock cycles than there are active transmitters. Thus, fairness is enforced, preventing starvation.

The simulation results for the three different arbitration methods appear in Fig. 7.9, which shows the average latency for bus sizes of 6 and 9 PEs using uniform traffic injection. Our exploration included a much wider range of bus sizes (i.e., number of bus clients), but the results invariably followed the same trends. Hence, only two bus sizes are shown here for illustrative purposes.

For all bus sizes, all three algorithms perform identically until the saturation throughput, λ_S, is reached, where FBF exhibits higher latency than both LWF and ST. This is most attributable to increasing starvation of PEs at higher injection rates in FBF, as previously discussed.

Similarly, the results for multimedia injection exhibited the same tendencies and, therefore, are not shown due to space considerations. The most notable difference in the latency response between uniform and multimedia traffic is the sharpness of the

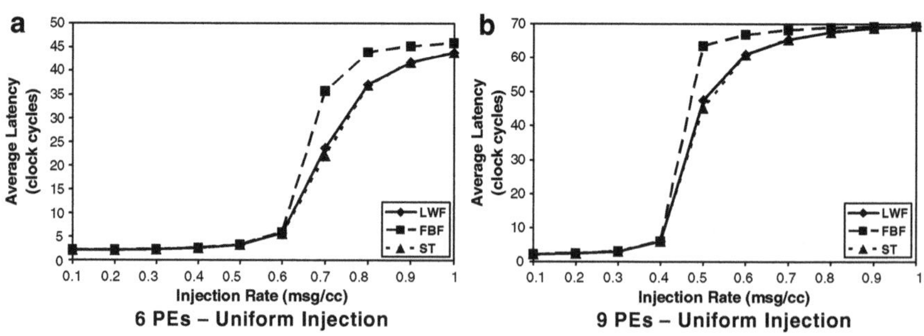

Fig. 7.9 Latency using three different arbitration policies. (**a**) 6 PEs – uniform injection. (**b**) 9 PEs – uniform injection

transition from non-saturated latency to saturation latency – multimedia injection transitions into saturation much more rapidly than uniform injection does.

The important and interesting result derived from these experiments is the high efficiency of the Sliding Timeslice arbitration policy. Even though its premise is very simple, the ST scheme consistently provides the best overall performance (or performs at least as well) when compared against the other two schemes. In fact, its only real competitor is the Longest Wait First method (see Fig. 7.9), but ST either matches or outperforms it slightly. Given that ST is significantly simpler to implement in hardware (since it does not need to maintain PE transmission order and wait-time states), it is the clear winner in this contest. This conclusion is testament to the fact that a simple solution may sometimes outwit more elaborate and expensive mechanisms.

7.3.4.4 Power Consumption

Power consumption was evaluated by implementing the dTDMA architecture in Verilog and synthesizing using 90 nm TSMC libraries (we did not have access to commercial libraries with smaller feature sizes). Power values for individual components, as reported by Synopsys Design Compiler, were imported into the dTDMA simulator. Based on actual component utilization, the total power consumption was calculated.

Figure 7.10 shows the power consumption of several dTDMA bus sizes over all uniform injection rates. Since the majority of power is consumed by the dynamic switching activity of the bus lines, interfaces, and arbiter, it comes as no surprise that power consumption behaves in the same manner as latency with respect to injection rate. Additionally, since dynamic power is proportional to $C_L * V_{dd}^2 * f$

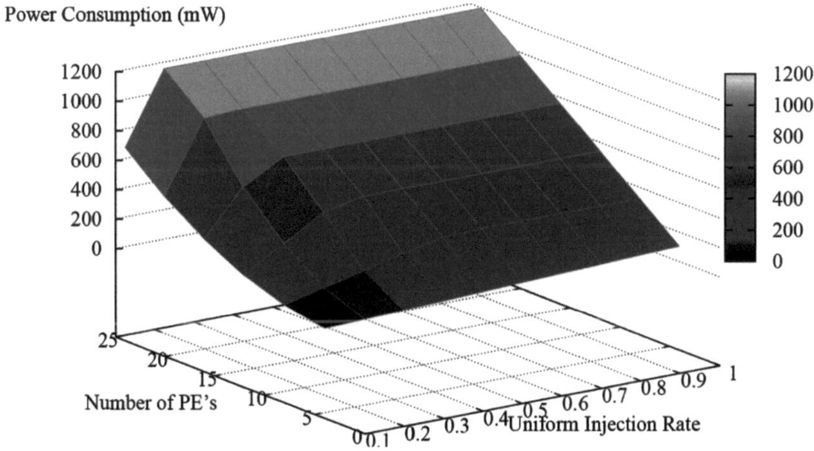

Fig. 7.10 Power consumption of the dTDMA bus

and C_L is linearly proportional to the length of the bus, dynamic power consumption grows linearly with bus size. The bus length, L_{BUS}, can be approximated as $L_{BUS} \approx \left\lceil \frac{P}{2} \right\rceil * L_P$, where P is the number of bus clients (PEs). This expression assumes that half the clients are placed on one side of the bus, with the other half placed on the other side. L_P is the length of the client (PE) exposed to (i.e., aligned with) the bus. Once L_{BUS} is estimated, a model like the Predictive Technology Model [146] can be used to extract the capacitance, C_L. The RC parasitics can also be extracted to determine the maximum frequency the bus can support (see Section 7.3.2).

Our synthesis results indicate that a dTDMA arbiter for a 9-PE bus consumes a mere 204.98 μW, while the receiver controller (i.e., *Rx* excluding the buffers) consumes 97.39 μW (the transmitter, *Tx*, is identical to *Rx*). In terms of absolute power values, 4- , 8- , and 16-PE dTDMA buses, at injection rate 1.0, consume about 106, 324, and 759 mW, respectively. Compare this to the AMBA bus, which consumes about 160, 232, and 368 mW for 4, 8, and 16 masters, respectively [148]. The increased power consumption of the dTDMA bus is attributable to its broadcast bus design (AMBA uses a multiplexed bus).

7.4 Comparison with Networks-on-Chip

One criterion for choosing a bus or an NoC is the number of PEs present in the system. Capacitive loading and contention – problems that limit the performance of large buses – are mitigated by an NoC interconnect. In this section, we compare the scalability of the dTDMA bus and NoCs.

7.4.1 Experimental Setup

A state-of-the-art NoC design appearing in [40] uses a link width of 128. A bus width of 512 bits was chosen for the dTDMA bus as the equivalent comparison for the following two reasons.

First, each NoC router has the ability to send *and* receive data on each of its four links (North, South, East, and West) at once. This means that with each clock cycle, a 128-bit NoC router can transmit 512 bits to its neighboring routers, and receive 512 bits from neighboring routers. Since the bus is inherently half-duplex, to mimic the NoC abilities we would need two 512-bit buses. However, this is excessive as the bus would be severely underutilized in low traffic conditions. It will be shown later that, even with a 256-bit bus, dTDMA significantly outperforms an NoC with 128-bit links.

Second, the dTDMA bus width can be increased without suffering decreased speed or increased complexity. On the contrary, increasing the bit width in an

NoC greatly affects three crucial architectural components within the node router: the input buffers, the virtual channel logic and buffers, and the switch crossbar. Specifically, an NoC cannot scale easily for the following reasons: (1) architectural components of each router will increase in size with bit width, (2) router components, in addition to increasing in size, will consume more power as bit width increases, and (3) each router is replicated n times in an n-PE mesh NoC. Using a Verilog implementation of an on-chip router implemented in 90 nm TSMC libraries, we found that doubling the bit width of the NoC incurred an area increase of 55% and a power increase of 77% per router. Additionally, since each router has five connections, each $2b$ bits wide (b bits in each direction), doubling the bit width incurred a tenfold increase in wiring complexity. The router in our work was modeled based on the low latency NoC router of [40].

A modified version of the simulator from Chapter 5 was used in conjunction with the dTDMA simulator described in the previous section to produce the results in the following sub-section. In the NoC simulator, routers have twelve virtual channels each with a 4-flit (one message) buffer, an input buffer of 32 flits (8 messages) from the local PE connection, and they employ wormhole switching. A *flit* is the smallest unit of flow control in NoC implementations – it typically corresponds to the bit width of the links (128 bits in this case). Accordingly, the dTDMA simulator uses an 8-message transmit buffer on each PE. A frequency of 500 MHz was used in both simulators to estimate power consumption. Injection rate is defined as the number of 512-bit messages injected per PE per clock cycle. Both simulators used a uniform random source-destination pattern. Similar to Section 7.3, two types of traffic injection were used – uniform and multimedia injection.

7.4.2 Results

Using the setup described above, experiments were carried out to find the average latency and power consumption of different size dTDMA buses and NoC meshes with respect to load rate.

With uniform injection (Fig. 7.11a–d), the dTDMA bus consistently outperforms the NoC mesh for PE populations of 4, 9, 16, and 25. NoC performance degrades more severely with injection rate than the dTDMA bus. However, at higher numbers of PEs, the dTDMA latency nears that of the NoC. These results indicate that dTDMA mitigates the contention issues, which afflict traditional bus architectures. Hence, dTDMA allows for up to 25 PEs to be placed on a bus without excessive contention problems. However, capacitive loading remains a limiting factor. It was shown in Section 7.3.2 that the maximum bus clock frequency drops below 500 MHz beyond 9 PEs. Thus, for designs operating at or below 500 MHz, dTDMA will still outperform traditional bus architectures and NoCs even for PE populations of more than 9. To take advantage of both low latencies and high clock

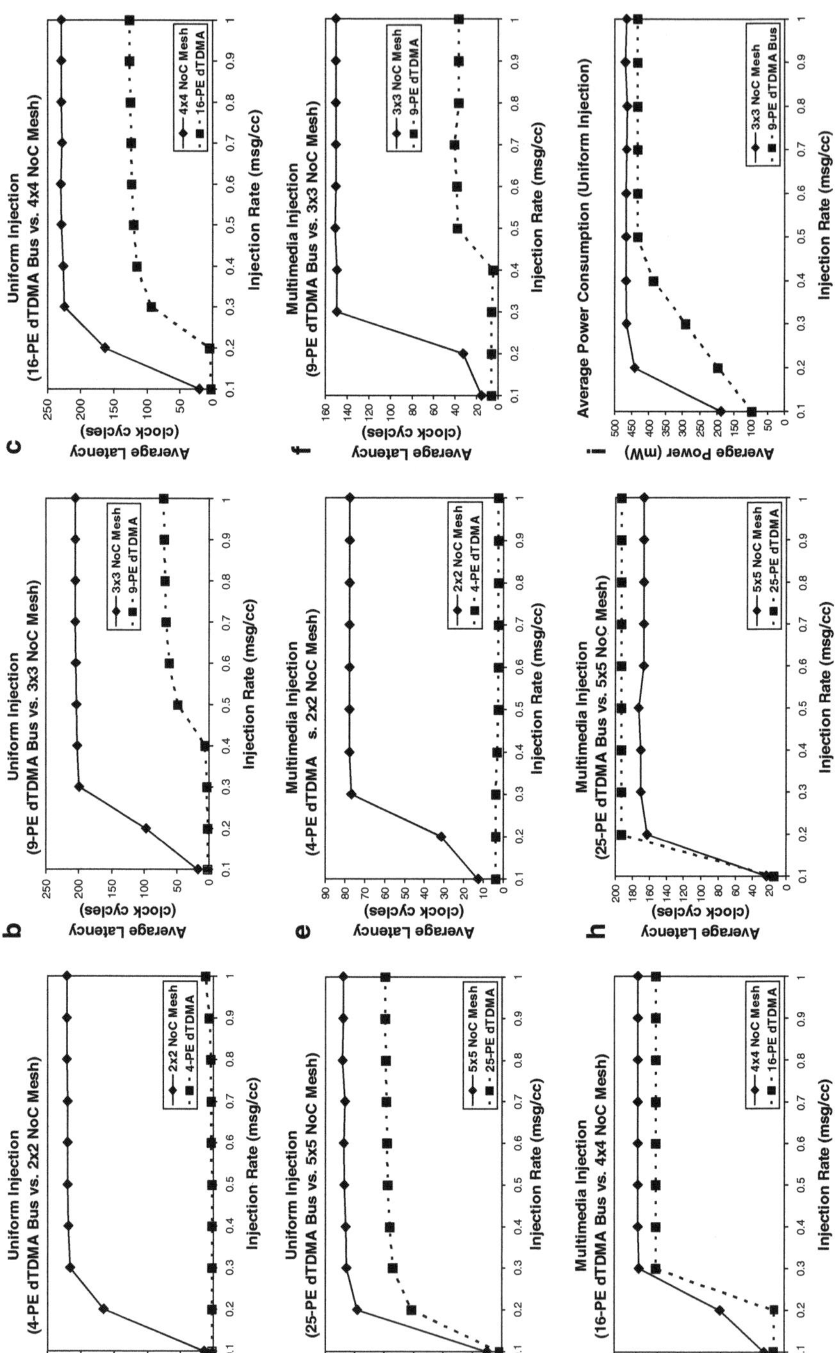

Fig. 7.11 dTDMA bus vs NoC latencies for uniform (**a–d**) and multimedia (**e–h**) traffic injection. A comparison of average power consumption is shown in (**i**)

frequencies, though, we conservatively conclude that the dTDMA bus performs optimally with PE populations of 9 and fewer. This result is in agreement with other evaluations of practical bus size [13]. Even at 9 PEs, the dTDMA bus shows a 66% reduction in latency over the NoC.

Multimedia traffic results, shown in Fig. 7.11e–h, exhibit trends similar to those of uniform traffic. A slight decrease in latency with increasing injection rate in (e) is visible because, at lower load rates, PEs do not transmit often enough to retain their timeslots and frequently must re-arbitrate to get a new timeslot. The same phenomenon, compounded with the large bus size, causes the 25 PE bus to perform worse than the NoC, as shown in Fig. 7.11h.

Results for a 256-bit bus are similar for 9 and fewer PEs. Though worse in absolute latency compared to the 512-bit bus by 32%, the results were still better than the NoC for all load rates. Thus, given equal *wiring area* (each side of the NoC router has two 128-bit links), and discounting the router area overhead of the NoC, the dTDMA bus still outperforms the NoC. Similarly, equal wiring area can be achieved by doubling the NoC bit width instead of halving the dTDMA bus width. Hence, each side of the NoC router would have two 256-bit links. This scenario was also simulated with the NoC showing a 25% improvement in performance. This NoC improvement was not enough to outperform the 512-bit dTDMA bus at any load rate.

Power consumption was evaluated through synthesized designs in 90 nm technology, as described in Section 7.3.4.4. The results are summarized in Fig. 7.11i. The dTDMA bus consumes significantly less power than the NoC before reaching its saturation throughput, at which point its power consumption nears that of the NoC. This behavior is due to the fact that a message in the NoC consumes power in each hop along its path, and that, even when idle, the NoC has significantly more resources leaking power (buffers, an arbiter, and a crossbar switch in each router). When the injection rate increases, the NoC power behavior remains the same, however, the dTDMA bus lines themselves start switching very frequently, consuming a significant amount of dynamic power.

7.5 Interconnect Hybridization

The results presented in Sections 7.3 and 7.4 show seemingly disparate interconnect choices – the dTDMA bus and the NoC. The bus exhibits low latency and high bandwidth efficiency for small systems (about 9 PEs), while the NoC exhibits good scalability for large systems. However, neither system alone provides for an efficient communication backbone for a large-scale system-on-chip, as the bus does not scale well, and the NoC does not adequately handle high injection rates.

Hybridization of the interconnect provides a solution to this problem. By taking advantage of the high speed, low latency bus for local communication, and by exploiting the NoC's scalability for global communication, an efficient system-wide interconnect can be created.

7.5.1 Affinity Grouping

Localized traffic congestion caused by traffic hotspots has been observed in existing network applications [82]. Such hotspots will increase the overall latency of the entire system because any message needing to pass through a hotspot will be delayed significantly. Adaptive routing is one solution to this problem, but it comes at the cost of increased router complexity due to the routing algorithm and deadlock prevention mechanisms.

Another means of reducing overall system latency is modification of the physical placement of PEs with respect to one another. Processing elements that communicate often with each other are said to exhibit communication *affinity*, and by placing PEs into affinity groups based on their communication patterns, the distance (hop count) traveled by a significant number of messages (those of the high-affinity PEs) can be reduced. The result is a lower delay for the PEs in the affinity group due to decreased hop count, and a lower system-wide latency due to decreased message traffic on network links outside of the affinity groups. Such grouping has been effectively utilized in SoC applications such as wireless decoding [149], multimedia [141], and cache-core interconnects for MPSoCs [19].

However, affinity grouping alone in a homogeneous network interconnect does not reduce local traffic load. In fact, grouping will increase the incidence of hotspots by creating areas of high congestion inside each affinity group. External traffic needing to pass through any one of these hotspots will incur undue delay. To isolate the traffic of an affinity group from the traffic of the rest of the system, a dedicated local interconnect can be used for the group. An interconnect that exhibits high throughput and low latency, but need not scale beyond several PEs, would be a good candidate. The dTDMA bus provides just such an interconnect. The local bus interconnect isolates the NoC from the intense communication occurring inside the affinity groups. This allows distant PEs to communicate over an NoC which is free of localized congestion. Moreover, the dTDMA bus provides for efficient multicasting – something that the NoC cannot provide without increasing the amount of network traffic or increasing the complexity of the routers [13].

Each affinity group is connected to the backbone NoC by a bridge PE. This bridge PE does no data processing; it merely acts as a relay, passing data back and forth between the local bus and the NoC. To send to a PE outside of its affinity group (foreign PE), a local PE uses its address lines to send its message to the bridge. The bridge inspects a header placed in front of the message data, and transmits the message as a series of packets on the appropriate NoC link according to the destination. The packet traverses the NoC, reaching the foreign bridge which inspects the message header and uses its address lines to send the message to the local recipient PE. Any PEs that are not able to be grouped (such as those which need to be accessed by all PEs) remain on the NoC.

7.5.2 Simulation Methodology

To examine the effects of hybridization, two interconnect architectures were compared with respect to average latency and power consumption. The first architecture, known as the Affinity Group (AG) Mesh, is an 8×8 mesh NoC. It contains four affinity groups of 8 PEs, but does not use the dTDMA bus for local interconnect inside the groups. The second architecture is the hybrid dTDMA/NoC, which is the AG Mesh converted to use the dTDMA bus as a local interconnect inside the groups. One bridge PE per affinity group connects to the NoC backbone. The dTDMA and NoC simulators used in the previous two sections were appropriately modified and integrated into a single simulation environment. This framework allows for seamless simulation of any bus/NoC hybrid topology.

Figure 7.12 shows how affinity grouping is applied to an NoC and how the affinity group is converted to use the dTDMA bus as a local interconnect. The 64-PE hybrid dTDMA/NoC topology was represented as a square mesh of dimension six. The four extra PEs in the 6×6 mesh were used to model the traffic injection of the affinity group bridges into the NoC, while the behavior of the 32 non-affinity group PEs remained the same as in the AG Mesh.

Different applications will exhibit varying degrees of affinity. For this reason, a parameter called the bridge rate was introduced. A bridge rate was defined for each PE, specifying how often the PE communicated with foreign PEs (PEs outside of the affinity group). A high bridge rate implies low affinity, and a low bridge rate implies high affinity. A similar technique, known as thread biasing, has been used to define the probability that threads executing on different cores will communicate with each other [19].

Bridge rates of 0.1–0.5 were simulated to model the dynamic nature of SoC traffic. While a bridge rate of 0.5 does indeed warrant the removal of a PE from an affinity group (it communicates within the affinity group as often as it communicates with other PEs), it is important to model the situation where the bridge rate occasionally rises to 50%. In the hybrid interconnect, each affinity group is

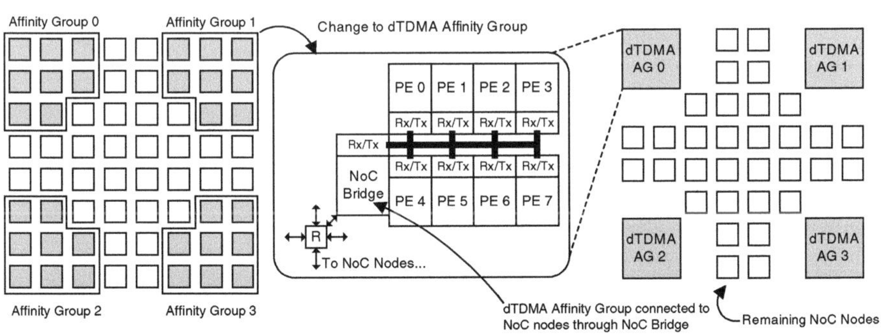

Fig. 7.12 Hybridization of the AG mesh using the dTDMA bus

represented by a single PE on the NoC. Each affinity group PE has an injection rate equal to the sum of the bridge rates of its constituent PEs.

In terms of source-destination selection, there are five possible message types in the hybrid network, namely: (1) AG-only (messages that originate and terminate within an affinity group), (2) AG-to-NoC (messages that originate in an affinity group and terminate at a non-AG PE in the NoC), (3) AG-to-AG (messages that originate in one affinity group and terminate in another), (4) NoC-to-AG (messages that originate in the NoC and terminate in an affinity group), and (5) NoC-only (messages that originate and terminate within the NoC).

As per Sections 7.3 and 7.4, uniform and multimedia traffic injection types were used. The combined dTDMA/NoC simulator reported the average system-wide latency and power consumption.

7.5.3 *Hybridization Results*

Figure 7.13 displays the results of various AG Mesh and dTDMA/NoC hybrid configurations. In all configurations, the dTDMA/NoC hybrid outperforms the AG Mesh by a sizeable margin. The worst-case latency reduction of 15.2% occurs with multimedia injection and a bridge rate of 0.1. For all bridge rates, the latency behavior is similar to that of the standalone dTDMA bus and NoC. However, saturation is reached more quickly with increasing injection rate as the system latency is now dominated by the NoC backbone, which saturates more quickly than the dTDMA bus.

As bridge rate is increased, overall system latency also increases. The cause of this is that there is an increasing amount of data traffic needing to pass outside of each affinity group, causing congestion on the NoC links on the periphery of the groups (in the case of the AG Mesh) or increasing contention for the bridge PE (in the case of the hybrid). A solution to this problem in the hybrid architecture is to use more than one bridge to connect each affinity group with the NoC backbone. To avoid complicated bridge arbitration, each PE in the group may be statically assigned a bridge to use. Mitigation of link congestion in the AG Mesh could be accomplished by adding direct links from the center of the affinity group to the NoC backbone. However, such *express lanes* do introduce additional wiring and routing overhead.

With regards to power consumption, the dTDMA/NoC hybrid consumes less power than the AG Mesh. This comes as no surprise since the hybrid architecture makes use of fewer NoC routers, in favor of the dTDMA bus hardware, which was shown to consume less power than the equivalent NoC architecture. Figure 7.14 shows the average power consumption of the AG Mesh and dTDMA/NoC hybrid over all bridge rates.

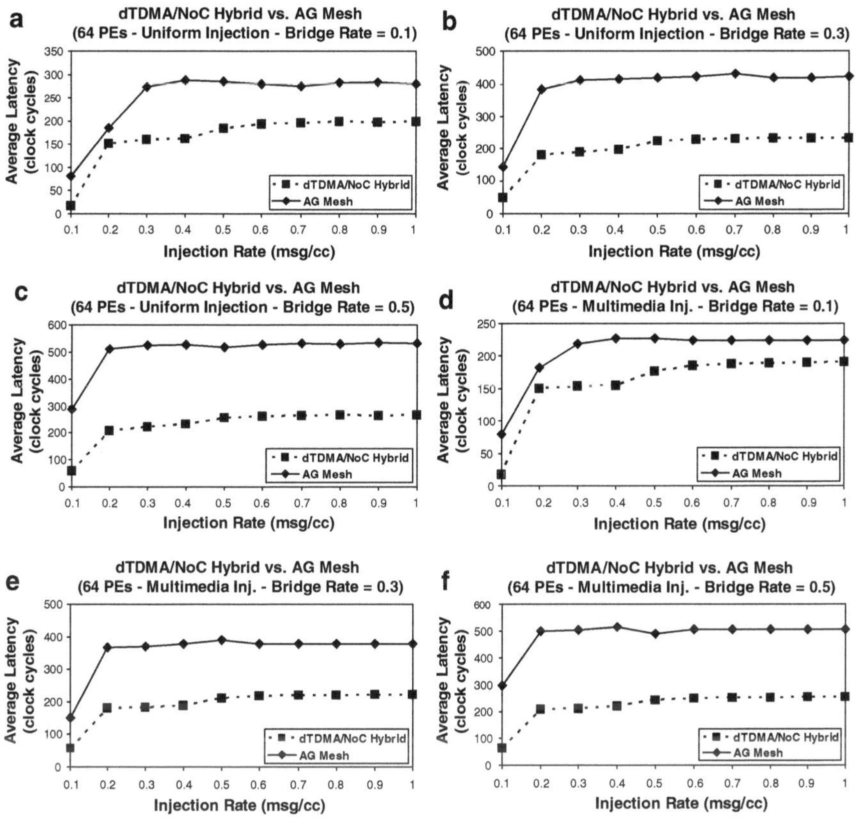

Fig. 7.13 Hybrid dTDMA bus/NoC vs pure NoC (AG mesh) latencies for uniform (**a**–**c**) and multimedia (**d**–**f**) traffic injection

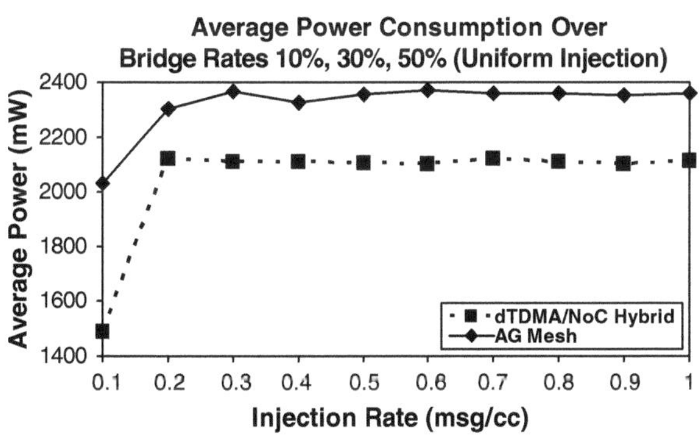

Fig. 7.14 Hybrid dTDMA bus/NoC vs pure NoC (AG mesh) average power consumption (uniform traffic injection)

7.6 Chapter Summary

The work presented in this chapter demonstrates the need for a new style of SoC interconnect. It has been shown that the master–slave, memory-mapped, transactional model of traditional bus architectures introduces several inefficiencies when applied to the SoC design space. Moreover, a preliminary analysis of CDMA interconnects concludes that three main obstacles, namely throughput reduction, codeword/sequence choice and management, and channel representation, will likely hinder the adoption of CDMA as a viable on-chip interconnect.

A new transaction-less bus was designed in response to the need for an efficient SoC interconnect. This bus, the dTDMA bus, uses a dynamic timeslice management system that results in near-100% bandwidth efficiency. The bus also natively supports multicasting and simplifies the implementation of various forms of QoS. When compared to traditional bus architectures, the dTDMA bus performs arbitrations significantly more quickly, resulting in lower latencies.

To support future large-scale SoCs, a hybrid interconnect system using affinity grouping and the dTDMA bus in conjunction with an NoC backbone was developed. By placing frequently-communicating PEs into affinity groups and connecting them locally with the high-speed dTDMA bus, the traffic load on the NoC backbone can be reduced.

Chapter 8
Design and Management of 3D Chip Multiprocessors Using Network-In-Memory (NetInMem) [43]

It has already been established that long interconnects are becoming an increasingly important problem from both power and performance perspectives. This motivates designers to adopt on-chip network-based communication infrastructures and three-dimensional (3D) designs where multiple device layers are stacked together.

Considering the current trends towards increasing use of chip multiprocessing, it is timely to consider 3D chip multiprocessor design and memory networking issues, especially in the context of data management in large L2 caches. The overall goal of this chapter is to study the challenges for L2 design and management in 3D chip multiprocessors.

The work presented hereafter constitutes a natural continuation of the previous chapter, which hybridized two seemingly disparate interconnection architectures – namely a bus and an on-chip network – to extract combined overall benefits. While Chapter 7 explored this concept in the traditional, 2D context, Chapter 8 will extend the exploration to the third dimension.

The recent surge toward 3D chip integration has been primarily fueled by two factors:

(1) The increasing number of on-die processing cores and caches – instigated by the explosion in transistor densities – has put enormous strain on off-chip memory bandwidth. Pad limited 2D chips simply cannot provide enough memory bandwidth to feed the barrage of processing elements on-chip. Consequently, designers looked in the third dimension to substantially increase the number of links arriving at a specific core. By adding potentially two additional gateways to the on-chip cores (one from above and one from below), architects can now envision systems with enough memory bandwidth to take advantage of the sprawling multi-core revolution. In such systems, one chip layer could consist of primarily memory cells sitting directly on top of a layer dominated by processing cores.

(2) Modern electronic systems contain mixed-signal technology, such as analog wireless transceivers integrated with a digital processing backbone. Traditionally, the analog and digital components were implemented in different chips and then combined on a single board. However, 3D chip integration can provide

C. Nicopoulos et al., *Network-on-Chip Architectures: A Holistic Design Exploration*, Lecture Notes in Electrical Engineering 45,
DOI 10.1007/978-90-481-3031-3_8, © Springer Science+Business Media B.V. 2009

significant area savings and boost performance by placing diverse technologies on separate layers of the same chip. The possibilities are, indeed, endless with technologies such as Micro-Electro-Mechanical Systems (MEMS) and emerging nanoscale non-CMOS solutions each occupying different layers of a 3D chip.

Figure 8.1 shows the progression from the 2D world of Chapter 7 to the 3D environment of Chapter 8.

The first contribution of this chapter is to propose a router architecture and a topology design that makes use of a network architecture embedded into the L2 cache memory. The second contribution is to demonstrate, through extensive experiments, that a 3D L2 memory architecture generates much better results than the conventional two-dimensional (2D) designs under different number of layers and vertical (inter-wafer) connections. In particular, experimental results show that a 3D architecture with no dynamic data migration generates better performance than a 2D architecture that employs data migration. This also helps reduce power consumption in L2 due to a reduced number of data movements.

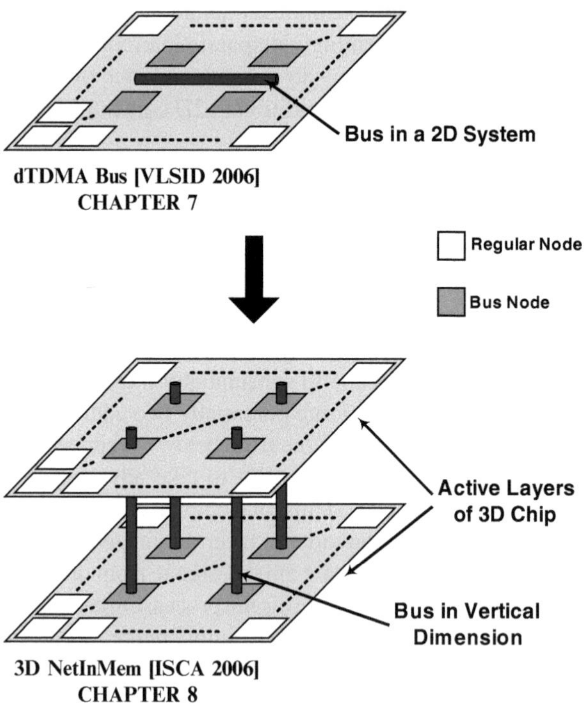

Fig. 8.1 The progression from 2D hybridization in Chapter 7 to 3D hybridization in Chapter 8

8.1 Introduction and Motivation

The adoption of Chip Multi-Processors (CMP) and other multi-core systems is expected to increase the size of both L2 and L3 caches in the foreseeable future. While traditional architectures have assumed that each level in the memory hierarchy has a single, uniform access time, increases in interconnect delays will render access times in large caches dependent on the physical location of the requested cache line. That is, access times will be transformed into variable latencies based on the distance traversed along the chip. The concept of Non-Uniform Cache Architectures (NUCA) [150] has been proposed based on the above observation. Instead of a large uniform monolithic L2 cache, the L2 space in NUCA is divided into multiple banks, which have different access latencies according to their location relative to the processor. These banks are connected through a mesh-based interconnection network. Cache lines are allowed to migrate within this network, for the purpose of placing more frequently-accessed data in the cache banks closer to the processor. Several recent proposals extend the NUCA concept to CMPs. An inherent problem of NUCA in CMP architectures is the management of data shared by multiple cores. Proposed solutions to this problem include data replication and data migration. Still, large access latencies and high power consumption stand as inherent problems for NUCA-based CMPs.

The introduction of three-dimensional (3D) circuits [151,152] provides an opportunity to reduce wire lengths. Consequently, this technology can be useful in reducing the access latencies to the remote cache banks of a NUCA architecture. In this chapter, we consider the design of a 3D topology for a NUCA that combines the benefits of network-on-chip and 3D technology to reduce L2 cache latencies in CMP-based systems. While network-on-chip and 3D cache designs have been studied in the past in different contexts, to our knowledge, this is the first in-depth study that integrates them. This chapter provides new insights on network topology design for 3D NoCs and addresses issues related to processor placement across 3D layers and data management in L2, taking into account network traffic and thermal issues. We evaluate the proposed architecture using a novel simulation environment that includes the Simics multi-core simulator [153] and a cycle-accurate 3D network simulator. Our experiments with the SPEC OMP benchmark suite [154] indicate that the proposed 3D architecture significantly reduces L2 access latencies over two-dimensional (2D) NUCA implementations (around 17 cycles on average), leading to an IPC improvement of up to 37%. In addition, our experiments show that a 3D architecture with no dynamic data migration generates better performance than a 2D architecture that employs data migration. This also helps reduce power consumption in L2 due to reduced data movement.

8.2 Background

In this sub-section, we first review previous schemes that exploit the non-uniform access patterns of large L2 caches. We then introduce the Network-In-Memory (NetInMem) and the three-dimensional architecture design issues.

8.2.1 NUCA Architectures

The issue of placement and location of data in large L2 caches has been identified by past research as one of the critical problems preventing us from extracting maximum performance from single- and multi-core machines. Recent proposals to large L2 design include decomposing L2 space into multiple, individually-addressable tiles (also called banks) whose access latencies depend on the distance between them and the processor that accesses them. Such architectures, usually known as Non-uniform Cache Architectures (NUCA), present unique challenges, as compared to uniform L2 architectures, in terms of data placement and management. Kim et al. [150] study cache line mapping and search strategies in the context of a NUCA architecture. Chishti et al. [155] propose NuRapid, a NUCA architecture that decouples tag placement from data placement. They later extended NuRapid to work within a CMP environment [156]. A unique contribution of this NuRapid architecture is that it employs a flexible data placement and replication-based management scheme to cut L2 access latencies. Zhang and Asanovic [157] propose a technique whereby copies of local primary cache victims are kept within the local L2 cache slices. Beckman and Wood [158] extend the NUCA concept to a multi-core setting by proposing a data migration scheme. Huh et al. [159] study the problem of how to partition an L2 NUCA to reduce interconnect traffic. Our L2 cache management policies benefit from both [158] and [156]; however, our L2 architecture is three-dimensional. Consequently, we tailored our data placement and migration policies considering multiple layers and the additional proximity provided by 3D. In addition, our results show that a 3D architecture can work very well even if we do not implement any data migration.

8.2.2 Network-In-Memory (NetInMem)

To facilitate the aforementioned variation in access latencies, the cache can no longer be a monolithic structure, since the large size would be detrimental to access time. Instead, it should be divided into self-contained banks which can be individually addressed. This would create a continuum of access times based on the location of each bank relative to the CPU. However, such a scheme demands a very efficient interconnection network to minimize total access time. The solution might come from a related branch in chip design, namely SoCs. Systems-on-Chip incorporate several homogeneous or heterogeneous processing elements on a single die. Since large NUCAs are expected to include tens of banks (as many as 256 have been seen in the literature), NoCs naturally emerge as the appropriate choice for the interconnect, similar to large SoCs. Cache banks would be tiled in a Manhattan-like grid, and nodes would be connected through a micro-network made up of a series of routers.

8.2.3 Three-Dimensional (3D) Design and Architectures

The idea of having a large number of L2 banks on a two-dimensional plane poses several challenges, even if one employs an NoC. The large chip area occupied by the memory elements necessitates the use of several NoC routers. A cache access request destined for a distant bank would have to traverse a large number of NoC routers. Note that, even when using state-of-the-art single stage routers, the communication delay can be substantial if the number of routers in a flit's path is large. It is, therefore, imperative to limit the number of routers encountered by a flit between its source and destination (hop count) in order to optimize the NUCA's performance. The up-and-coming 3D chip design space is particularly amenable to reducing the Manhattan distance (and thus hop count) in large chips.

A three dimensional (3D) chip is a stack of multiple device layers with direct vertical interconnects tunneling through them [151,152]. The benefits of 3D ICs include: (1) higher packing density due to the addition of a third dimension to the conventional two-dimensional layout, (2) higher performance due to reduced average interconnect length, and (3) lower interconnect power consumption due to the reduction in total wiring length [161]. Joyner et al. [161] have shown that three-dimensional architectures reduce wiring length by a factor of the square root of the number of layers used. For example, a 4-layer 3D NoC would have, on average, $\approx \sqrt{4} = 2$ times shorter wiring length, as illustrated in Fig. 8.2.

There are currently various 3D technologies being explored in industry and academia, but the two most promising ones are Wafer-Bonding [151] and Multi-Layer Buried Structures (MLBS) [162]. Wafer-bonding technology processes each active device layer separately and then connects the layers in a single entity. For MLBS, the front-end processing is repeated on a single wafer to build multiple device layers, before the back-end process builds interconnects among the devices. A trait of critical importance in these technologies is the size of the 3D vias which

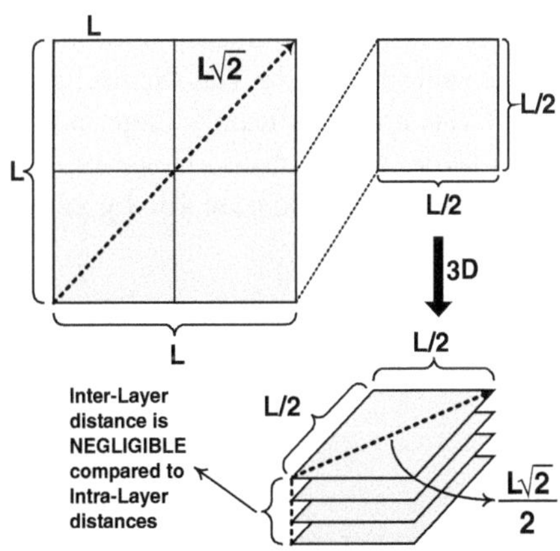

Fig. 8.2 Wiring scales in length as the square root of the number of layers in three dimensions

connect neighboring layers together. In wafer-bonding, the size of the 3D vias is not expected to scale at the same rate as feature sizes [163]. MLBS, on the other hand, can provide vias that scale down with feature size. However, since MLBS is not compatible with current manufacturing processes, it is not as appealing as wafer bonding techniques [164,165]. Various 3D integration vertical interconnect technologies have been explored, including wire bonded, micro-bump, contact-less (capacitive or inductive), and through-via vertical interconnect [166]. Through-via interconnection has the potential to offer the greatest vertical interconnect density and therefore is the most promising one among these vertical interconnect technologies. There are two different approaches to implementing through-via 3D integration: the first one involves sequential device process, in which the front-end processing (to build the device layer) is repeated on a single wafer to build multiple active device layers, before the interconnects among devices are built. The second approach processes each active device layer separately, using conventional fabrication techniques, and then stacking these multiple device layers together using wafer-bonding technology. The latter approach requires minimal changes to the manufacturing steps and is more promising; therefore, it is adopted in our proposed architecture. Furthermore, there are currently two primary wafer orientation schemes, Face-To-Face [164] and Face-To-Back [160,165], as shown in Fig. 8.3. While the former provides the greatest layer-to-layer via density, it is suitable for two-layer organizations, since additional layers would have to employ back-to-back placement using larger and longer vias. Face-To-Back, on the other hand, provides uniform scalability to an arbitrary number of layers, despite a reduced inter-layer via density. Hence, to provide scalability and easy manufacturability, we assume in this work the use of Face-To-Back Wafer-Bonding.

One critical issue which has emerged in the design of 3D chips is the inter-layer via pitch, which dictates the density of layer-to-layer interconnections. Via pitches can range from $1\times1\ \mu m^2$ to $10\times10\ \mu m^2$ [151], depending on the technology and manufacturing process used. Very recently, IBM has managed to reduce the pitch to a state-of-the-art $0.2\times0.2\ \mu m^2$ using Silicon-On-Insulator (SOI) technology [167]. However, despite the decreasing via sizes, it is the via pads (i.e., the via endpoints) which ultimately limit the via density. Currently, via pads do not scale at the same rate as the vias themselves. Compared to a wire pitch of 0.1 μm, inter-layer vias are significantly larger and cannot achieve the same wiring density as intra-layer interconnects. However, what vertical interconnects lack in terms of density, they compensate for by extremely small inter-wafer distances, ranging

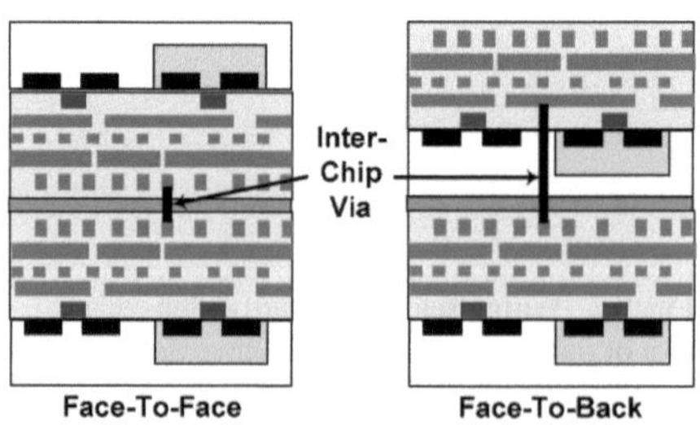

Fig. 8.3 Face-to-face and face-to-back wafer-bonding in 3D technology [160]

from 5 to 50 μm. Such distances are extremely small, providing rapid transfer times when traversing layers [168].

Thermal considerations have been a significant concern for 3D integration [169]. However, various techniques have been developed to address thermal issues in 3D architectures such as physical design optimization through intelligent placement [170], increasing thermal conductivity of the stack through insertion of thermal vias [169], and use of novel cooling structures [171]. Further, a recent work demonstrated that the areal power density is the more important design constraint in placement of the processing cores in a 3D chip, as compared to their location in the 3D stack [172]. Consequently, thermal concern can be managed as long as components with high power density are not stacked on top of each other. Architectures that stack memory on top of processor cores, or those that rely on low-power processor cores have been demonstrated to not pose severe thermal problems [173]. In spite of all these advances, one can anticipate some increase in temperature as compared to a 2D design, and also a temperature gradient across layers. Increased temperatures increase wire resistances, and consequently the interconnect delays. To capture this effect, we study the impact of temperature variations on the 3D interconnect delay to assess the effect on performance.

Researchers have so far focused on physical aspects, low-level process technologies, and developing automated design and placement tools [151,160,174]. Research at the architectural level has also surfaced [163,164]. Specific to 3D memory design, [168,175] have studied multi-bank uniform cache structures. In our work, we focus our attention on the design of a 3D NoC-based non-uniform L2 cache architecture.

Modern System-on-Chip (SoC) designs, such as CMPs, can benefit from 3D integration as well. For example, by placing processing memory, such as DRAM and/or L2 caches, on top of the processing core in different layers, the bandwidth between them can be significantly increased and the critical path can be shortened [176]. In [177], a CMP design with stacked memory layers is proposed. The authors show that the L2 cache can be removed due to the availability of wide low-latency inter-layer buses between the processing core layer and DRAM layers, and the area saved from this can be recycled for additional cores. The notion of adding specialized system analysis hardware on separate active layers stacked vertically on the processor die using 3D IC technology is explored in [178]. The modular snap-on introspective layer collects system statistics and acts like a hardware system monitor.

8.3 A 3D NetInMem Architecture

Our proposed architecture for multiprocessor systems with large shared L2 caches involves placement of CPUs on several layers of a 3D chip with the remaining space filled with L2 cache banks. Most 3D IC designs observed in the literature so far have not exceeded five layers, mostly due to manufacturability issues, thermal management, and cost. A detailed analysis of via pitch considerations and their impact on the number of inter-layer gateways follows in Section 8.3.1. As previously mentioned, the most valuable attribute of 3D chips is the very small distance between the layers. A distance on the order of tens of microns is negligible compared to the distance traveled between two network on-chip routers in 2D (1,500 μm on average for a 64 KB cache bank implemented in 70 nm technology). This characteristic makes traveling in the vertical (inter-layer) direction very fast as compared to the horizontal (intra-layer).

One inter-layer interconnect option is to extend the NoC into three dimensions. This requires the addition of two more links (up and down) to each router. However, adding two extra links to an NoC router will increase its complexity (from five links to seven links). This, in turn, will increase the blocking probability inside the router since there are more input links contending for an output link. Moreover, the NoC is, by nature, a multi-hop communication fabric, thus it would be unwise to place traditional NoC routers on the vertical path because the multi-hop delay and the delay of the router itself would overshadow the ultra fast propagation time.

It is not only desirable, but also feasible, to have single-hop communication amongst the layers because of the short distance between them. To that effect, we propose the use of dynamic Time-Division Multiple Access (dTDMA) buses (see Chapter 7) as "Communication Pillars" between the wafers, as shown in Fig. 8.4.

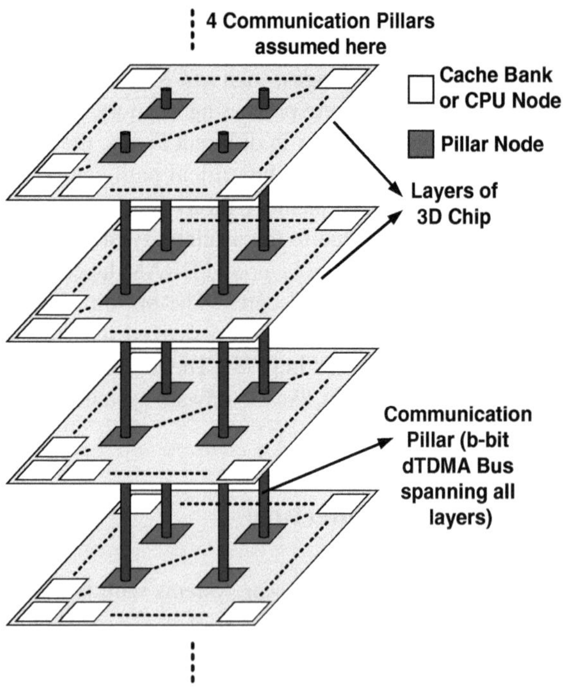

Fig. 8.4 The proposed 3D Network-In-Memory (NetInMem) architecture

These vertical bus pillars provide single-hop communication between any two layers, and can be interfaced to a traditional NoC router for intra-layer traversal using minimal hardware, as will be shown later. Furthermore, hybridization of the NoC router with the bus requires only one additional link (instead of two) on the NoC router. This is the case because the bus is a single entity for communicating both up and down. Due to technological limitations and router complexity issues (to be discussed in Section 8.3.1), not all NoC routers can include a vertical bus, but the ones that do form gateways to the other layers. Therefore, those routers connected to vertical buses have a slightly modified architecture, as explained in Section 8.3.2.

8.3.1 The dTDMA Bus as a Communication Pillar

The dTDMA bus architecture (Chapter 7 and [15]) eliminates the transactional character commonly associated with buses, and instead employs a bus arbiter which dynamically grows and shrinks the number of timeslots to match the number of active clients. Single-hop communication and transaction-less arbitrations allow for low and predictable latencies. Dynamic allocation always produces the most efficient timeslot configuration, making the dTDMA bus nearly 100% bandwidth efficient. Each pillar node requires a compact transceiver module to interface with the bus, as shown in Fig. 8.5.

As already explained in the previous chapter, the dTDMA bus interface (Fig. 8.5) consists of a transmitter and a receiver connected to the bus through a tri-state driver. The tri-state drivers on each receiver and transmitter are controlled by independently programmed fully-tapped feedback shift registers. Details of its operation can be found in [15]. The total number of wires required by the control signals from the arbiter to each layer is $3n+\log_2(n)$, for n layers. Because of its very small size, the dTDMA bus interface is a minimal addition to the NoC router.

The presence of a centralized arbiter is another reason why the number of vertical buses, or pillars, in the chip should be kept low. An arbiter is required for each

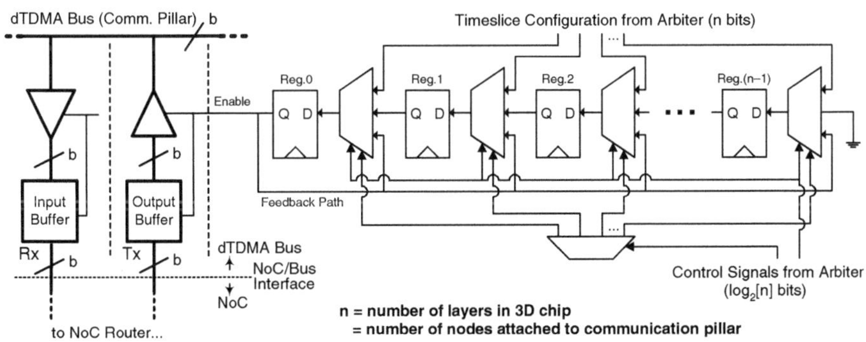

Fig. 8.5 Transceiver module of a dTDMA bus

Fig. 8.6 Side view of the 3D chip with the dTDMA bus (pillar)

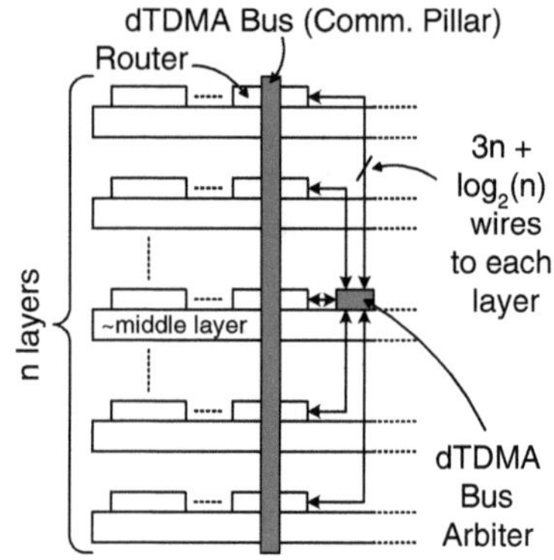

Table 8.1 Area and power overhead of dTDMA bus

Component	Power	Area
Generic NoC router(5-port)	119.55 mW	0.3748 mm^2
dTDMA bus Rx/Tx (2 per client)	97.39 μW	0.00036207 mm^2
dTDMA bus arbiter (1 per bus)	204.98 μW	0.00065480 mm^2

pillar with control signals connecting to all layers, as shown in Fig. 8.6. The arbiter should be placed in the middle layer of the chip to keep wire distances as uniform as possible. Naturally, the number of control wires increases with the number of pillar nodes attached to the pillar, i.e., the number of layers present in the chip. The arbiter and all the other components of the dTDMA bus architecture have been implemented in Verilog HDL and synthesized using commercial 90 nm TSMC libraries. The area occupied by the arbiter and the transceivers is much smaller compared to the NoC router, thus fully justifying our decision to use this scheme as the vertical gateway between the layers. The area and power numbers of the dTDMA components and a generic 5-port (North, South, East, West, local node) NoC router (all synthesized in 90 nm technology) are shown in Table 8.1. Clearly, both the area and power overheads due to the addition of the dTDMA components are orders of magnitude smaller than the overall budget. Therefore, using the dTDMA bus as the vertical interconnect is of minimal area and power impact. A 7-port NoC router was considered and eliminated in the design search due to prohibitive contention issues, multi-hop communication in the vertical direction, and substantially increased area/power overhead due to an enlarged crossbar and more complicated switch arbiters. The dTDMA bus was observed to be better than an NoC for the vertical direction as long as the number of device layers was less than 9 (bus contention becomes an issue beyond that).

The length of vertical interconnect between two layers is assumed to be 10 μm. According to [166], the parasitics of inter-tier vias have a small effect on power and delay, because of their small length (i.e., low capacitance) and large cross-sectional area (i.e., low resistance).

The density of the inter-layer vias determines the number of pillars which can be employed. Table 8.2 illustrates the area occupied by a pillar consisting of 170 wires

Table 8.2 Area overhead of inter-wafer wiring for different via pitch sizes

	Inter wafer area(due to dtbma bus wiring)			
Bus width	10 μm	5 μm	1 μm	0.2 μm
128 bits(+42 control)	62500μm²	15625μm²	625μm²	25μm²

(128-bit bus + 3 × 14 control wires required in a 4-layer 3D SoC) for different via pitch sizes. In Face-To-Back 3D implementations (Fig. 8.3), the pillars must pass through the active device layer [168], implying that the area occupied by the pillar translates into wasted device area. This is the reason why the number of inter-layer connections must be kept to a minimum. However, as via density increases, the area occupied by the pillars becomes smaller, and, at the state-of-the-art via pitch of 0.2 μm, becomes negligible compared to the area occupied by the NoC router (see Tables 8.1 and 8.2). However, as previously mentioned, via densities are still limited by via pad sizes, which are not scaling as fast as the actual via sizes. As shown in Table 8.2, even at a pitch of 5 μm, a pillar induces an area overhead of around 4% to the generic 5-port NoC router, which is not overwhelming. These results indicate that, for the purposes of our 3D architecture, adding extra dTDMA bus pillars is feasible.

Via density, however, is not the only factor limiting the number of pillars. Router complexity also plays a key role. As previously mentioned, adding an extra vertical link (dTDMA bus) to an NoC router will increase the number of ports from 5 to 6, and since contention probability within each router is directly proportional to the number of competing ports, an increase in the number of ports increases the contention probability. This, in turn, will increase congestion within the router, since more flits will be arbitrating for access to the router's crossbar. Thus, arbitrarily adding vertical pillars to the NoC routers adversely affects the performance of each pillar router. Hence, the number of high-contention routers (pillar routers) in the network increases, thereby increasing the latency of both intra-layer and interlayer communication.

On the other hand, there is a minimum acceptable number of pillars. In this work, we place each CPU on its own pillar. If multiple CPUs were allowed to share the same pillar, there would be fewer pillars, but such an organization would give rise to other issues, as described in the following sub-section.

8.3.2 CPU Placement

In our proposed architecture, we use a single-stage router to minimize latency. Routers connected to pillar nodes are different, as an interface between the dTDMA pillar and the NoC router must be provided to enable seamless integration of the vertical links with the 2D network within the layers. The modified router is shown in Fig. 8.7. An extra physical channel (PC) is added to the router, which corresponds to the vertical link. The extra PC has its own dedicated buffers, and is indistinguishable from the other links to the router operation. The router only sees an additional physical channel.

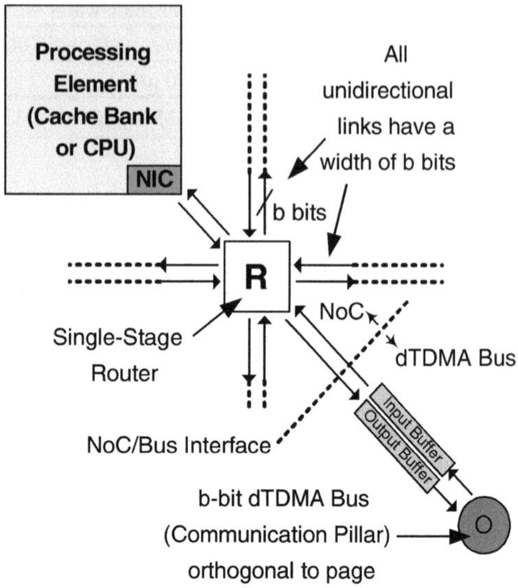

Fig. 8.7 A high-level overview of the modified router of the pillar nodes

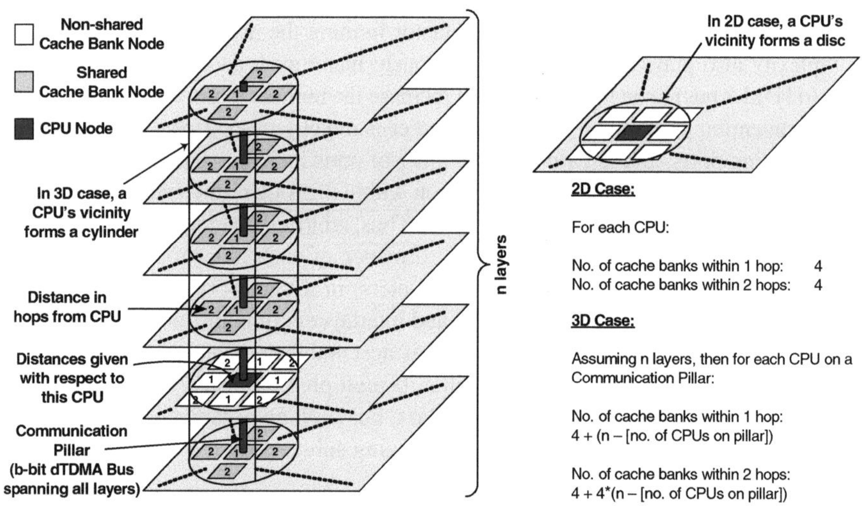

Fig. 8.8 A CPU has more cache banks in its vicinity in the 3D architecture

The dTDMA pillars provide rapid communication between layers of the chip. We have a dedicated pillar associated with each processor to provide fast inter-layer access, as shown in Fig. 8.4. Such a configuration gives each processor instant access to the pillar, additionally providing them with rapid access to all cache banks that are adjacent to the pillar. By placing each processor directly on a pillar, its memory locality (the number of banks with low access latency) is increased in the vertical direction (potentially both above and below), in addition to the pre-existing

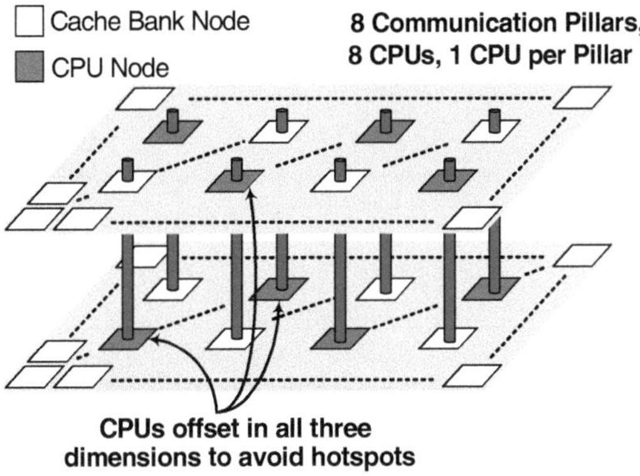

Fig. 8.9 Hotspots can be avoided by offsetting CPUs in all three dimensions

locality in the 2D plane. This is illustrated in Fig. 8.8. Such an increase in the number of cache banks with low access latency can significantly improve the performance of applications. The relative sizing of L2 cache banks and CPU+L1 cache, as shown in Fig. 8.8, is meant to be illustrative. Our placement approach works even when a CPU+L1 cache span the size of multiple L2 cache banks.

Stacking CPUs directly on top of each other would give rise to thermal issues. Increased temperatures due to layer stacking are a major challenge in 3D design [164], and are often a major determining factor in component placement. Since the CPUs are expected to consume the overwhelming majority of power (they are constantly active, unlike the cache banks), it would be thermally-unwise to stack any two or more processors in the same vertical plane. Furthermore, stacking processors directly on top of each other on the same pillar would affect the performance of the network as well, as it would create high congestion on the pillar. Processors are the elements which generate most of the L2 traffic (there is also some traffic generated by the migration algorithm, as explained in Section 8.4.2); therefore, forcing them to share a single link would create excessive traffic. Our simulations in later sections will validate this argument. To avoid thermal and congestion problems, CPUs can be offset in all three dimensions (maximal offsetting), as shown in Fig. 8.9.

If, however, the manufacturing technology provides only low via densities, the designer might be forced to use fewer pillars than the number of CPU cores to minimize the wasted device-layer area. In such cases, multiple CPUs would have to share a single pillar, warranting a careful CPU placement methodology, to ensure maximum performance with minimum thermal side-effects. The best way to achieve this is to offset the processors at various distances from the pillar, while keeping them within the direct vicinity. Moving the processors far away from the pillars would be detrimental to performance, since changing layers would inflict lengthy 2D traversal times to and from the pillar. In our approach, the CPUs are placed according to the pattern illustrated in Fig. 8.10. The processors are placed at most two hops away from a pillar to minimize the adverse effect on performance.

To perform the placement of the CPUs based on the restrictions of Fig. 8.10, a simple algorithm was developed to achieve uniform offset of all CPUs in both the

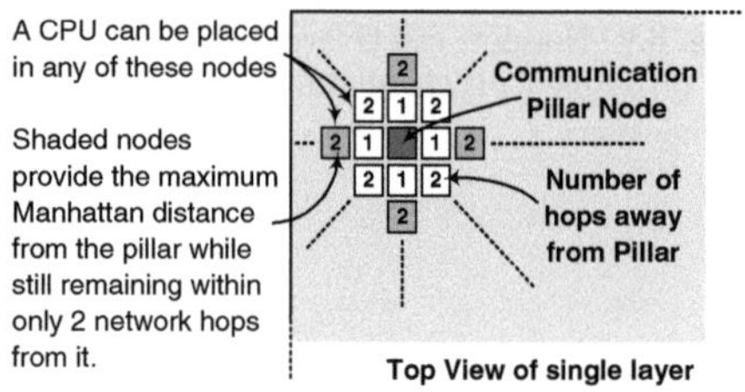

Fig. 8.10 Placement pattern of CPUs around the pillars

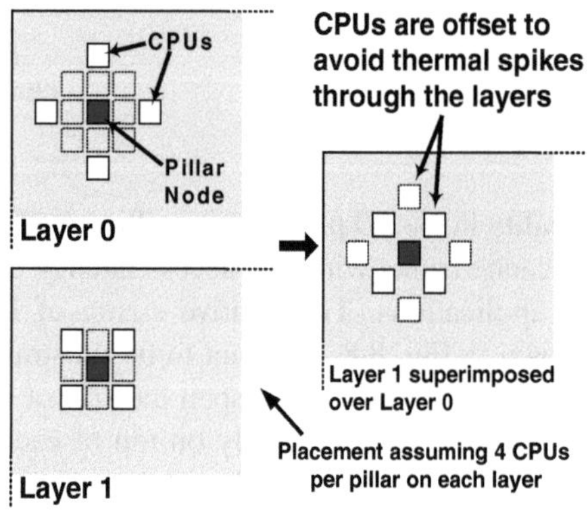

Fig. 8.11 An example of the CPU placement algorithm

vertical and horizontal directions for the configurations studied. The algorithm, shown in Algorithm 1, assumes placement of 2 or 4 CPUs per pillar per chip layer. The parameter k is the offset distance from a pillar in number of network hops. In our proposed implementation, k was chosen to be 1. If thermal issues require more mitigation, then k can be increased at the expense of performance. Parameter c is the number of CPUs assigned to each pillar on each layer. Assigning more than 4 CPUs per pillar per layer is not desirable, since it will increase bus contention dramatically, thus degrading network performance. The location of the pillars is assumed to be predetermined by the designer, and is given as a constant to the algorithm. The pillars need to be placed as far apart from each other as possible within the layer to avoid the creation of network congested areas. On the other hand, the pillars should not be placed on the edges because such placement would limit the number of cache banks which are in the vicinity of the pillar.

The placement pattern spans four layers, beyond which it is repeated. This is done because (1) increasing the factor of k beyond 2 moves the CPU too far from the pillar and (2) thermal effects reduce as inter-layer distance between processors increases. Note that the placement is scalable to any number of layers.

To visualize how the algorithm works, Fig. 8.11 shows an example for four CPUs per pillar per layer. The key observation is that this placement methodology provides a systematic and scalable way to place the CPUs in a thermal aware fashion.

To validate the CPU placement methodology proposed above, we simulated the thermal profile of several configurations using HS3d, a 3D thermal estimation tool [179].

c: number of CPUs per pillar
k: distance offset from pillar in terms of the number of network hops
for $i \in Pillars$ {
  for $l \in Layers$ {
    //x_i,y_i of each pillar assumed predetermined
    pillar node $i = (x_i, y_i, l)$
    switch l modulo 4 {
    case 0: if $c == 2$ then
          CPU's at $(x_i \pm k, y_i, l)$
        else if $c == 4$ then
          CPU's at $(x_i \pm 2k, y_i, l)$ and $(x_i, y_i \pm 2k, l)$
    case 1: if $c == 2$ then
          CPU's at $(x_i, y_i \pm k, l)$
        else if $c == 4$ then
          CPU's at $(x_i \pm k, y_i \pm k, l)$
    case 2: if $c == 2$ then
          CPU's at $(x_i \pm 2k, y_i, l)$
        else if $c == 4$ then
          CPU's at $(x_i \pm k, y_i, l)$ and $(x_i, y_i \pm k, l)$
    case 3: if $c == 2$ then
          CPU's at $(x_i, y_i \pm 2k, l)$
        else if $c == 4$ then
          CPU's at $(x_i \pm 2k, y_i \pm 2k, l)$
    }
  }
}

Algorithm 1. CPU placement algorithm for more than one CPU/pillar/layer

HS3d provides estimates for peak, minimum and average temperatures of a chip, based on the power consumption and location of all architectural components. The tool also provides a detailed steady-state thermal profile for the entire floor plan, allowing us to identify hotspots. The power consumption of the cache banks was extracted from Cacti 3.2 [180]. Regarding the CPU cores, we assumed a simple, single-issue core, similar to recent trends in industry (STI Cell Processor, Sun UltraSPARC T1) [3,4]. Sun's UltraSPARC T1 (formerly codenamed Niagara) employs 8 processing cores with peak power consumption of 79 W [181]. Using this information as a guideline for future CMPs, we approximate the power consumption per core to be 8W in our thermal simulation (assuming that the rest of the power is consumed in peripheral circuits and L2 cache). Since our experiments focus on the relative trends between the 2D and 3D topologies, the results will still hold even if this absolute power consumption is not accurate.

The results of the thermal simulations on a system with 256 64KB L2 cache banks (16MB) and 8 CPU cores are shown in Table 8.3. The L2 cache banks are clock-gated when not in use. As expected, moving from a 2D layout to a 3D layout causes the average temperature of the chip to increase. However, of critical importance is the avoidance of hotspots, i.e., places where the peak temperature is extremely high. Hotspots can substantially degrade performance and adversely affect the lifetime of the chip. Hotspots can be avoided through careful offsetting of the processor cores. Offsetting the cores in all three dimensions provides the best

Table 8.3 Temperature profile of various configurations (*L* is the number of layers and *k* is the offset factor in Algorithm 1)

Configuration	Peak temp °C	Avg temp °C	Min temp °C
2D, maximal offset	111.05	53.96	46.77
3D-2L, optimal offset	119.05	63.94	49.21
3D-2L, offset K=2	125.02	63.94	49.59
3D-2L, offset K=1	135.24	63.94	49.52
3D-2L, CPU stacking	173.38	63.94	50.73
3D-4L, optimal offset	158.67	86.62	64.79
3D-4L, CPU stacking	287.12	86.62	58.51

results, causing an increase in peak temperature of only 8°C when using two layers, as compared to the 2D case. Assuming a single CPU per pillar (row 2 of Table 8.3), CPUs can be optimally offset in all three dimensions, as in Fig. 8.9. However, when processors need to share pillars, then we use the offsetting technique of Algorithm 1. As shown in Table 8.3, increasing the offset factor, *k*, can reduce peak temperature by 10°C. As predicted, stacking CPUs in both 2-layer and 4-layer configurations is detrimental to peak temperature (i.e., leads to hotspot creation).

8.4 3D L2 Cache Management

In this sub-section, we present our organization of processor and L2 cache banks, and then detail our L2 cache management policies.

8.4.1 Processors and L2 Cache Organization

Figure 8.12 illustrates the organization of the processors and L2 caches in our design. Similar to CMP-DNUCA [158], we separate cache banks into multiple clusters. Each cluster contains a set of cache banks and a separate tag array for all the cache lines within the cluster. Some clusters have processors placed in the middle of them, while others do not. All the banks in a cluster are connected through a network-on-chip for data communication, while the tag array has a direct connection to the local processor in the cluster. Note that even though it is not explicitly shown in Fig. 8.12, each processor has its own private L1 cache and an associated tag array for L2 cache banks within its local cluster. For a cluster without a local processor, the tag array is connected to a customized logic block which is responsible for receiving a cache line request, searching the tag array and forwarding the request to the target cache bank. This organization of processors and caches can be scaled by changing the size and/or number of the clusters.

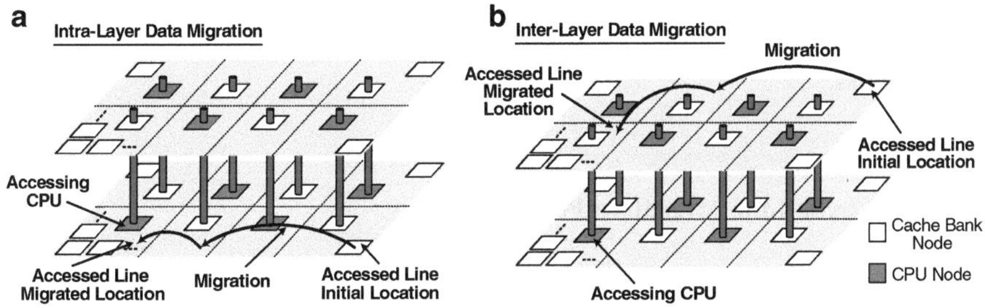

Fig. 8.12 Intra-layer and inter-layer data migration in the 3D L2 architecture (*dotted lines* denote clusters)

8.4.2 Cache Management Policies

Based on the organization of processors and caches given in the previous sub-section, we developed our cache management policies, consisting of a cache line search policy, a cache placement and replacement policy, and a cache line migration policy, all of which are detailed in the following sub-sections.

8.4.2.1 Search Policy

Our cache line search strategy is a two-step process. In the first step, the processor searches the local tag array in the cluster to which it belongs and also sends requests to search the tag array of its neighboring clusters. All the vertically neighboring clusters receive the tag that is broadcast through the pillar. If the cache line is not found in either of these places, then the processor multicasts the requests to the remaining clusters. If the tag match fails in all the clusters, then it is considered as an L2 miss. On a tag match in any of the clusters, the corresponding data is routed to the requesting processor through the network-on-chip.

8.4.2.2 Placement and Replacement Policy

We use cache placement and replacement policies similar to those of CMP-DNUCA [158]. Initially a cache line is placed according to the low-order bits of its cache tag, that is, these bits determine the cluster in which the cache line will be placed initially. The low-order bits of the cache index indicate the bank in the cluster into which the cache line will be placed. The remaining bits of the cache index determine the location in the cache bank. The tag entry of the cluster is also updated when the cache line is placed. The placement policy can only be used to determine the initial location of a cache line as when cache lines start migrating, the lower order bits of the cache tag can no longer indicate the cluster location. Finally, we use a pseudo-LRU replacement policy to evict a cache line to service a cache miss.

8.4.2.3 Cache Line Migration Policy

Similar to prior approaches, our strategy attempts to migrate data closer to the accessing processor. However, our policy is tailored to the 3D architecture and migrations are treated differently based on whether the accessed data lies in the same or different layer as the accessing processor. For data located within the same layer, the data is migrated gradually to a cluster closer to the accessing processor. When moving the cache lines to a closer cluster, we skip clusters that have processors (other than the accessing processor) placed in them since we do not want to affect their local L2 access patterns and get the cache lines to the next closest cluster without a processor. Eventually, if the data is accessed repeatedly by only a single processor, it migrates to the local cluster of the processor. Figure 8.12a illustrates this intra-layer data migration.

For data located in a different layer, the data is migrated gradually closer to the pillar closest to the accessing processor (see Fig. 8.12b). Since clusters accessible through the vertical pillar communications are considered to be in local vicinity, we never migrate the data across the layers. This decision has the benefit of reducing the frequency of cache line migrations, which in turn reduces power consumption.

To avoid false misses (misses caused by searches for data in the process of migration), we employ a lazy migration mechanism as in CMP-DNUCA [158].

8.5 Experimental Evaluation

8.5.1 Methodology

We simulated the 3D CMP architecture by using Simics [153] interfaced with a 3D NoC simulator. A full-system simulation of an 8-processor CMP architecture running Solaris 9 was performed. Each processor uses in-order issue and executes the SPARC ISA. The processors have private L1 caches and share a large L2 cache. The default configuration parameters for processors, memories and Network-In-Memory are given in Table 8.4. Some of the parameters in this table are modified for studying different configurations. The shown cache bank and tag array access latencies are extracted using Cacti 3.2 [180].

To model the latency of the three-dimensional, hybrid NoC/bus interconnect, we developed a cycle-accurate simulator in C, based on an existing 2D NoC simulator [41]. For this work, the 2D simulator was extended to three dimensions,

Table 8.4 Default system configuration parameters (L2 cache is organized as 16 clusters of size 16×64 KB)

Process parameters	
Number of process	8
Issue width	1
Memory parameters	
L1(split i/d)	64 KB, 2-way, 64 B line, 3-cycle, write-through
L2 (unified)	16 MB (256×64 KB), 16-way, 64 B line, 5-cycle bank access
Tag array (per cluster)	24 KB, 4-cycle access
Memory	4 GB, 260 cycle latency
Network parameters	
Number of layers	2
Number of pillars	8
Routing scheme	Dimension-order
Switching scheme	Wormhole
Flit size	128 bits
Router latency	1 cycle

and the dTDMA bus was integrated as the vertical communication channel. The 3D NoC simulator produces, as output, the communication latency for cache access.

In our cache model, private L1 caches of different processors are maintained coherent by implementing a distributed directory-based protocol. Each processor has a directory tracking the states of the cache lines within its L1 cache. L1 access events (such as read misses) cause state transitions and updates to directories, based on the MSI protocol. The traffic due to L1 cache coherence is taken into account in our simulation.

We simulated nine SPEC OMP benchmarks [154] with our simulation platform. These benchmarks are listed in Table 8.5 (we could not run the remaining two SPEC OMP benchmarks through Simics due to a memory leak problem). For each benchmark, we marked an initialization phase in the source code. The cache model is not simulated until this initialization completes. This is reflected as the fast-forward cycles for each benchmark shown in the second row of Table 8.5. After that, each application runs 500 million cycles for warming up the L2 caches. We then collected statistics for the next 2 billion cycles following the cache warm-up period. The third row in Table 8.5 gives the total number of L2 cache accesses (including data read, data write, and instruction fetch) within the sampling period for each

Table 8.5 Benchmarks used

Benchmarks	Ammp	Apsi	Art	Equake	Fma3d	Galgel	Mgrid	Swim	Wupwise
Fast-forward (Million cycles)	3,633	4,453	3,523	21,538	18,535	3,665	3,533	4,306	18,777
Number of L2 tran-sactions	24,508,715	27,013,477	25,638,435	27,502,906	12,599,496	38,181,613	204,815,737	164,762,040	141,499,738

benchmark. We see that the benchmarks mgrid, swim and wupwise exhibit many more L2 accesses than the others, as a result of higher L1 miss rates.

8.5.2 Results

We first introduce the schemes compared in our experiments. We refer to the scheme with perfect search from [158] as CMP-DNUCA. We name our 2D and 3D schemes as CMP-DNUCA-2D and CMP-DNUCA-3D, respectively. Note that our 2D scheme is just a special case of our 3D scheme discussed in the chapter, with a single layer. Both of these schemes employ cache line migration. To isolate the benefits due to 3D technology, we also implemented our 3D scheme without cache line migration, which is called CMPSNUCA-3D.

Our first set of results give the average L2 hit latency numbers under different schemes. The results are presented in Fig. 8.13. We observe that our 2D scheme (CMPDNUCA-2D) generates competitive results with the prior 2D approach (CMP-DNUCA [158]). Our 2D scheme shows slightly better IPC results for several benchmarks because we place processors not on the edges of the chip, as in CMPDNUCA, but instead surround them with cache banks as shown in Fig. 8.12. Our results with 3D schemes reiterate the expected benefits from the increase in locality. It is interesting to note that CMP-SNUCA-3D, which does not employ migration, still outperforms the 2D schemes that employ migration. On the average, L2 cache latency reduces by 10 cycles when we move from CMP-DNUCA-2D to CMP-SNUCA-3D. Further gains are also possible in the 3D topology using data migration. Specifically, CMPDNUCA-3D reduces average L2 latency by 7 cycles

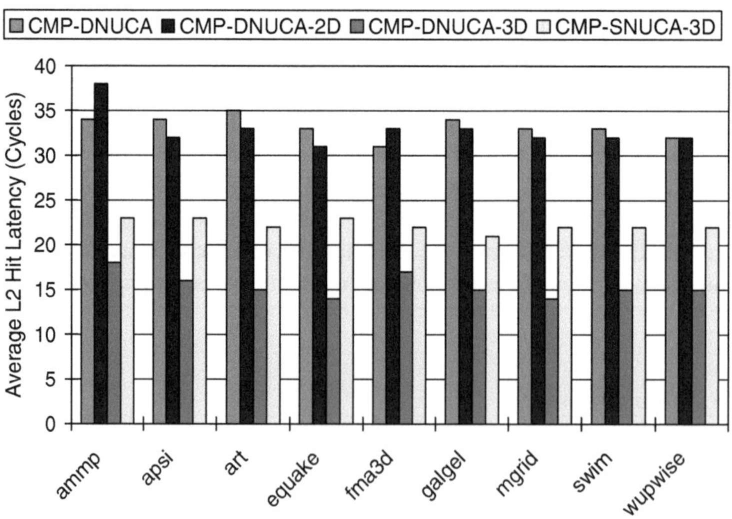

Fig. 8.13 Average L2 hit latency values under different schemes

as compared to the static 3D scheme. Further, we note that even when employing migration, as shown in Fig. 8.14, 3D exercises it much less frequently compared to 2D, due to the increased locality (see Fig. 8.8). The reduced number of migrations in turn reduces the traffic on the network and the power consumption. These L2 latency savings translate to IPC improvements commensurate with the number of L2 accesses. Figure 8.15 illustrates that the IPC improvements brought by CMP-DNUCA-3D (CMP-SNUCA-3D) over our 2D scheme are up to 37.1% (18.0%). The IPC improvements are higher with mgrid, swim and wupwise since these applications exhibit higher number of L2 accesses.

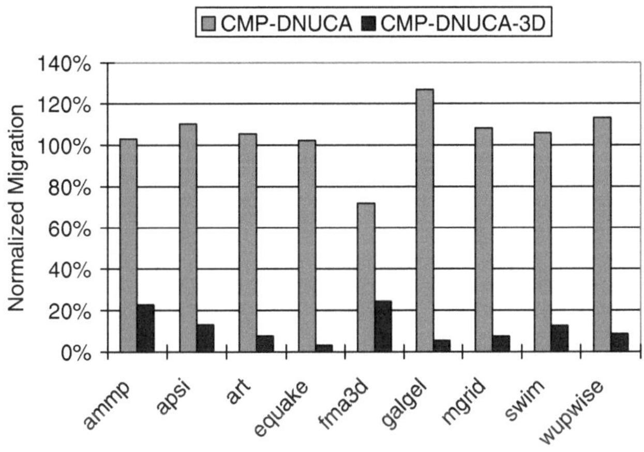

Fig. 8.14 Number of block migrations for CMP-DNUCA and CMP-DNUCA-3D, normalized with respect to CMP-DNUCA-2D

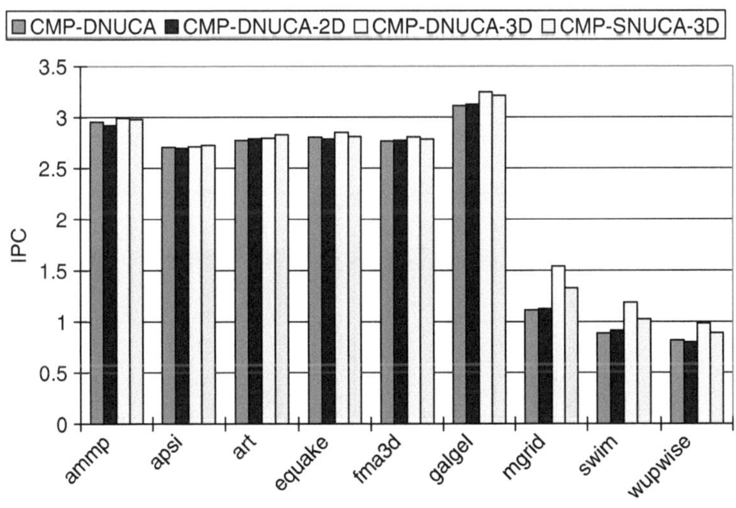

Fig. 8.15 IPC values under different schemes

We then proceed to study the impact of larger cache sizes on our savings using CMP-DNUCA-2D and CMP-DNUCA-3D. When we increase the size of the L2 cache, we increase the size of each cluster, while maintaining the 16-way associativity. Figure 8.16 shows the average L2 latency results with 32MB and 64MB L2 caches for four representative benchmarks (art and galgel with low L1 miss rates and mgrid and swim with high L1 miss rates). We observe that L2 latencies increase with the large cache sizes albeit at a slower rate with the 3D configuration (on average 7 cycles for 2D versus 5 cycles for 3D), indicating that 3D topology is a more scalable option when we move to larger L2 sizes.

Next, we conduct experiments by modifying some of the parameters in the underlying 3D topology. The results with the CMP-DNUCA-3D scheme using different numbers of pillars to capture the effect of the different interlayer via pitches are given in Fig. 8.17. As the number of pillars reduces, the contention for

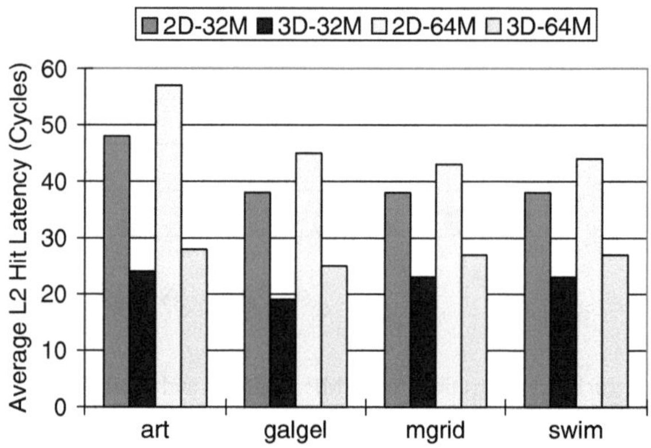

Fig. 8.16 Average L2 hit latency values under different schemes

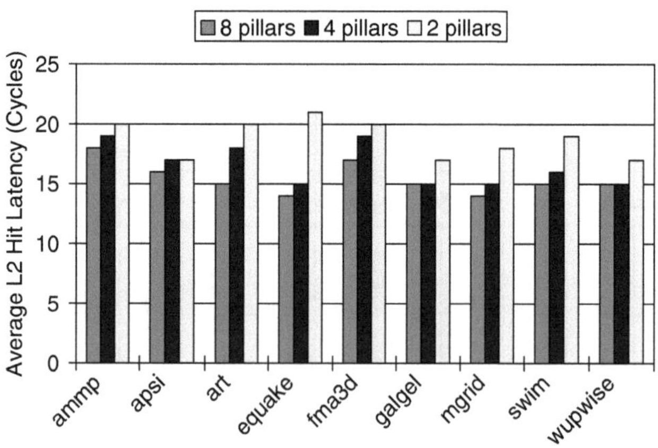

Fig. 8.17 Impact of the number of pillars (CMP-DNUCA-3D scheme)

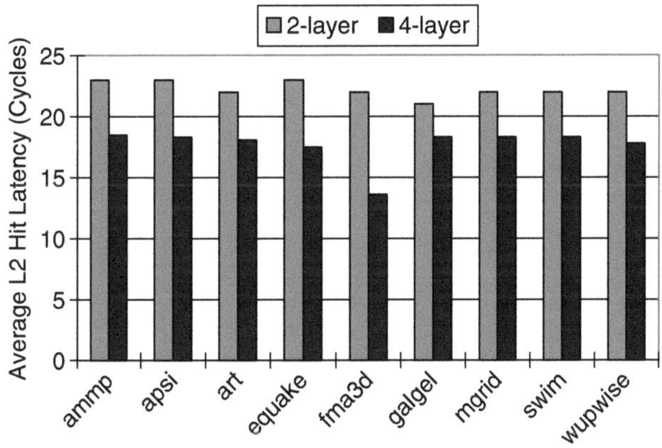

Fig. 8.18 Impact of the number of layers (CMP-SNUCA-3D Scheme)

the shared resource (pillar) increases to service interlayer communications. Consequently, average L2 latency increases by 1 to 7 cycles when we move from 8 to 2 pillars. Also, when the number of layers increases from 2 to 4, the L2 latency decreases by 3 to 8 cycles, primarily due to the reduced distances in accessing data, as illustrated in Fig. 8.18 for the CMP-SNUCA-3D scheme. However, it needs to be recalled that the additional layers impose a higher thermal cost as has been shown in Table 8.3.

8.6 Chapter Summary

Three dimensional circuits and Networks-on-Chip (NoC) are two emerging trends for mitigating the growing complexity of interconnects. In this chapter, we have demonstrated that combining on-chip networks and 3D architectures can be a promising option for designing large L2 cache memories for chip multiprocessors. Specifically, the author has proposed a novel hybrid bus/NoC fabric to efficiently exploit the fast vertical interconnects in 3D circuits, discussed processor placement and L2 data management issues, and presented an extensive experimental evaluation of the proposed architecture as well as its comparison to 2D L2 cache designs. Experiments have been performed using a novel simulation framework that integrates a 3D NoC simulator for L2 caches with a system level multi-CPU simulator. Experiments have shown that the proposed 3D architecture reduces average L2 access latency significantly over 2D topologies and this, in turn, brings IPC benefits. In addition, results in this work have shown that moving from a 2D topology to a 3D topology can provide more latency reduction than incorporating sophisticated data migration strategies into the 2D topology. It has also been demonstrated that the placement of processors needs to be done with care, taking into account the thermal issues. Furthermore, it has been shown that the bandwidth of the vertical interconnections (captured by varying the number of pillars) has a significant impact on the L2 cache latencies. Overall, the results of this chapter emphasize the importance of considering 3D technology in designing future chip multiprocessors.

Chapter 9
A Novel Dimensionally-Decomposed Router for On-Chip Communication in 3D Architectures [44]

Much like multi-storey buildings in densely packed metropolises, three-dimensional (3D) chip structures are envisioned as a viable solution to skyrocketing transistor densities and burgeoning die sizes in multi-core architectures. As already seen in the previous chapter, partitioning a larger die into smaller segments and then stacking them in a 3D fashion can significantly reduce latency and energy consumption. Such benefits emanate from the notion that inter-wafer distances are negligible compared to intra-wafer distances. This attribute substantially reduces global wiring length in 3D chips.

The work in this chapter expands on the previous chapter by further integrating the increasingly popular idea of packet-based Networks-on-Chip into a 3D setting. While NoCs have been studied extensively in the 2D realm, the microarchitectural ramifications of moving into the third dimension have yet to be fully explored. This chapter presents a detailed exploration of inter-strata communication architectures in 3D NoCs.

Chapter 8 investigated the notion of NoC/Bus hybridization to take advantage of the gross asymmetry between horizontal (i.e., intra-layer) and vertical (i.e., inter-layer) distances in 3D chips. By employing a bus structure for vertical transfers, single-hop communication between any two layers becomes possible. However, bus structures are limited by an inherent attribute they possess: their shared-medium pedigree. Since only one client can transmit at any one time on a bus, concurrent communication between different sets of transmitters/receivers is not possible, even if the communicating pairs are physically situated in non-overlapping portions of the bus. It is precisely this realization that led to the work presented in this chapter, i.e., investigating 3D inter-layer communication in great detail so as to develop a more efficient interconnect architecture.

Toward that extent, three design options are investigated: the simple bus-based inter-wafer connection of Chapter 8, a hop-by-hop standard 3D design, and a full 3D crossbar implementation. Figure 9.1 abstracts the progression from a hop-by-hop design to a segmented-link implementation in a conceptual manner. The goal is to overcome the shared-medium nature of the bus by providing segmented links that can accommodate seamless and concurrent communication between the different strata of a 3D chip. Furthermore, the NoC/Bus hybridization scheme of the previous chapter relied on 2D crossbars interconnected with a vertical bus. This chapter studies

C. Nicopoulos et al., *Network-on-Chip Architectures: A Holistic Design Exploration*, Lecture Notes in Electrical Engineering 45,
DOI 10.1007/978-90-481-3031-3_9, © Springer Science+Business Media B.V. 2009

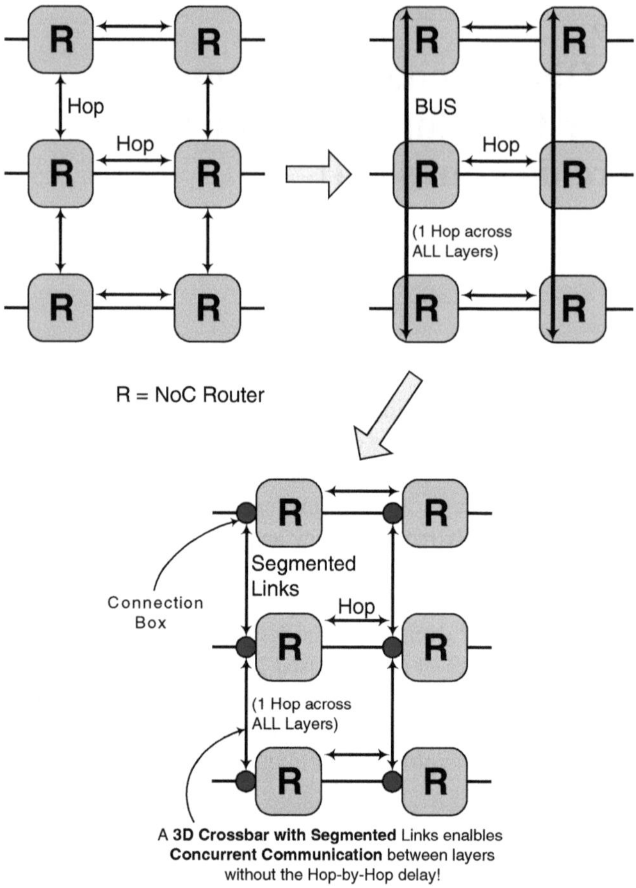

Fig. 9.1 The evolution from a hop-by-hop 3D structure to a true 3D crossbar

the feasibility of a true physical 3D crossbar that spans all routers in the same vertical cross-section, fusing them into a single operational entity. This allows for coordinated and concurrent inter-layer data transfers.

In this context, the author proposes a novel partially-connected 3D crossbar structure, called the 3D Dimensionally-Decomposed (DimDe) Router, which provides a good tradeoff between circuit complexity and performance benefits. Simulation results using (a) a stand-alone cycle-accurate 3D NoC simulator running synthetic workloads, and (b) a hybrid 3D NoC/cache simulation environment running real commercial and scientific benchmarks, indicate that the proposed DimDe design provides latency and throughput improvements of over 20% on average over the other 3D architectures, while remaining within 5% of the full 3D crossbar performance. Furthermore, based on synthesized hardware implementations in 90 nm technology, the DimDe architecture outperforms all other designs – including the full 3D crossbar – by an average of 26% in terms of the Energy-Delay Product (EDP).

9.1 Introduction and Motivation

Three-dimensional chip technology promises to reduce interconnect delays by stacking multiple layers on top of each other, and by providing shorter vertical connections [175]. This technology has matured and demystified some of the concerns on thermal viability and reliability of inter-wafer vias. In addition, it promises to enable integration of heterogeneous technologies on the same chip – such as having layers of memory stacked on top of processor cores – and is even attractive for placing analog and digital components on the same chip, as this avoids common substrate noise problems. Interconnect architecture design across the layers in a 3D architecture requires careful attention for the components on different layers to communicate effectively. Furthermore, there is a need for an integrated approach to interconnect design in the 2D planes and the vertical direction. Currently, there exists no systematic effort at exploring the interconnect architecture for 3D chips. Recently, researchers have started examining some tradeoffs, such as the influence of bandwidth variation of inter-layer interconnects between processor and memory subsystems [177], and – as described in the previous chapter – combining vertical interconnects with an NoC fabric for chip multiprocessor caches [43]. In this work, we investigate various architectural options for 3D NoC design. Interconnect design in 3D chips imposes new constraints and opportunities compared to that of 2D NoC design. There is an inherent asymmetry in the delays in a 3D architecture between the fast vertical interconnects and the horizontal interconnects that connect neighboring cores, due to differences in wire lengths (few tens of μm in the vertical direction as compared to few thousand μm in the horizontal direction). Consequently, extending a traditional NoC fabric to the third dimension by simply adding routers at each layer (called the Symmetric NoC in this work, due to the symmetry of routing in all directions) is not a good option, as router latencies may dominate the fast vertical interconnect. Hence, we explore two alternate options, a 3D NoC-bus hybrid structure and a true 3D router fabric for the vertical, inter-strata interconnect. A key challenge with 3D NoC routers is limiting the arbitration complexity due to the large path diversity resulting from the additional interconnects in the third dimension.

Vertical interconnects also impose a larger area overhead than corresponding horizontal wires due to the requirement for bonding pads, and can compete with device area as the inter-strata vias punch through the wafer when Face-to-Back (F2B) bonding (see Fig. 9.2) is used. Therefore, the desired number of vertical interconnects used in the 3D router architecture needs to be investigated. In exploring these tradeoffs in a 3D router design, we developed a new 3D NoC router architecture that we call the 3D Dimensionally-Decomposed (DimDe) Router. The name is a direct corollary of the fact that communication flow through the DimDe router is classified according to the three axes in Euclidean space: X (corresponding to East–West intra-layer traffic), Y (corresponding to North–South intra-layer traffic), and Z (corresponding to inter-layer traffic in the vertical dimension). The idea of decomposing traffic in two dimensions in a 2D environment was introduced in

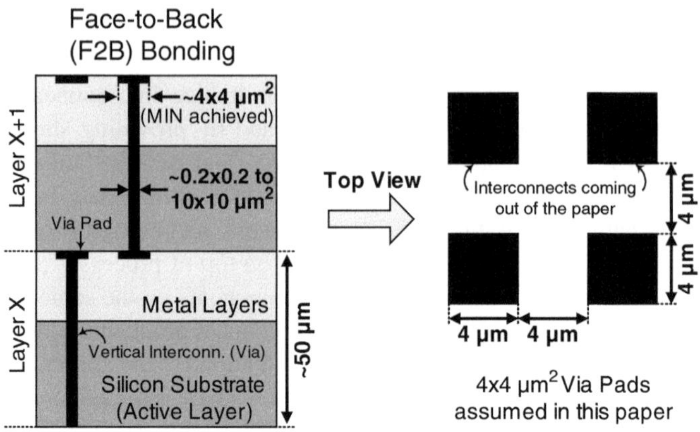

Fig. 9.2 Face-to-back (F2B) bonding and the assumed vertical via layout in Chapter 9

[182,183] and revisited more recently in the work described in Chapter 4 and [40]. While our proposed DimDe router was inspired by the concept presented in [40], our contribution goes well beyond the introduction of a new traffic dimension. *The DimDe router fuses the crossbars of all the routers in the same vertical "column" (i.e., same X, Y coordinate but different Z coordinate) into a unified entity that allows coordinated concurrent communication across different layers through the same crossbar*. This design amounts to a *true physical* 3D crossbar (unlike the mere stacking of 2D routers in multiple wafer layers). It is important to note that 3D topologies have long been in existence in the macro-network field (e.g., k-ary n-cube), but these rely on 2D routers connected in such a way as to form a *logical* 3D topology. However, 3D chip integration is now enabling the creation of a true physical 3D topology, where the router is itself a three-dimensional entity. DimDe exhibits the following characteristics that make it a desirable interconnect structure for 3D designs:

(1) DimDe supports a true 3D crossbar structure which spans all the active layers of the chip. Irrespective of the number of layers used in the implementation, the 3D crossbar allows a single-hop connection between any two layers, treating all strata as part of a single router structure.
(2) The DimDe design-space provides options for varying the number of vertical connections from one to four to emulate anything between a segmented bus and a full crossbar. Through design space exploration, DimDe was selected to support two vertical interconnects to strike a balance between the path diversity and high bandwidth offered by a full 3D crossbar and the simplicity of a bus. Most importantly, DimDe's partially-connected crossbar achieves performance levels similar to those of a full 3D crossbar, with substantially reduced area and power overhead and orders of magnitude lower control logic complexity.
(3) DimDe supports segmented vertical (i.e., inter-strata) links in the partially-connected crossbar to enable concurrent communication between the different layers of

the 3D chip. This simultaneous data transfer in the vertical dimension significantly increases the vertical bandwidth of the chip as compared to a 3D NoC-bus hybrid structure.

(4) The DimDe design employs a hierarchical arbitration scheme for inter-strata transfers that reduces area and delay complexity, while still efficiently enabling simultaneous data transfers. The first stage arbitrates between all requests for vertical communication from within a single layer and the second stage accommodates as many simultaneous requests from the winners of the first stage arbitration.

(5) Similar to the Row-Column (RoCo) Decoupled Router of [40], DimDe completely separates East–West and North–South intra-layer traffic through a pre-sorting operation at the input. However, interlayer traffic cannot be completely isolated in its own module. A true 3D crossbar requires inter-layer traffic to merge with intralayer traffic in a seamless fashion; this would allow incoming packets from different layers to continue traversal in the destination layer. DimDe facilitates this tight integration by augmenting the Row (East–West) and Column (North–South) modules with a Vertical Module which fuses with the other two. The Vertical Module then extends to all other layers and unifies them in a single operational entity. The Vertical Module assumes the double role of "gluing" all the layers together and blending inter- and intra-layer traffic through unidirectional connections to the Row and Column modules of all layers. It will be demonstrated that this approach dramatically reduces the 3D crossbar complexity, while still allowing concurrent communication between different layers through the switch.

The proposed 3D router design is compared to four different interconnect architectures: a 2D NoC, a 3D Symmetric NoC, a 3D NoC-Bus Hybrid, and a Full 3D Crossbar implementation (Our interpretation of a "full" 3D crossbar is presented in Section 9.2.3 and subsequently formalized in Section 9.4.3). To provide as comprehensive an evaluation as possible, we employed a two-pronged simulation environment: (a) a stand-alone, cycle-accurate NoC simulator running synthetic workloads, and (b) a hybrid NoC/cache simulator running a variety of commercial and scientific workloads within the context of a shared, multi-bank NUCA L2 cache in an 8-CPU Chip Multi-Processor (CMP) scenario. This double-faceted evaluation process ensures exposure to several traffic patterns, including request/reply memory traffic.

The proposed DimDe design consistently provides the lowest latency for different traffic patterns and it saturates at higher workloads compared to other considered architectures. Synthetic workload results show that, for high traffic loads, the recently proposed 3D NoC-Bus Hybrid Architecture of Chapter 8 and [43] exhibits the worst latency and throughput for all traffic patterns (even worse than the 2D topology), as the bus saturates first with higher workload. In terms of throughput behavior, the DimDe architecture provides 18% average improvement over the other designs, while remaining within around 3% of the Full 3D Crossbar's throughput. The real workload results indicate that DimDe provides an average improvement of 27% over the 3D Symmetric and 3D NoC-Bus Hybrid designs, and remains within 4% of the Full 3D Crossbar's performance. However, with the Energy-Delay Product (EDP) as the metric, *DimDe significantly outperforms all other designs, including the Full 3D Crossbar, by 26% on average*. Hence, when accounting for both performance and power consumption, the DimDe design is superior to all other 3D router architectures analyzed in this chapter. To the best of the author's knowledge, this is the first systematic exploration and analysis of 3D interconnect architectures and their ramifications on overall system performance.

9.2 Three-Dimensional Network-on-Chip Architectures

This section delves into the exploration of possible architectural frameworks for a three-dimensional NoC network. As described in Chapter 2, a typical 2D NoC consists of a number of Processing Elements (PE) arranged in a grid-like mesh structure, much like a Manhattan grid. The PEs are interconnected through an underlying packet-based network fabric. Each PE interfaces to a network router through a Network Interface Controller (NIC). Each router is, in turn, connected to four adjacent routers, one in each cardinal direction. Expanding this two-dimensional paradigm into the third dimension poses interesting design challenges. Given that on-chip networks are severely constrained in terms of area and power resources, while at the same time they are expected to provide ultra-low latency, the key issue is to identify a reasonable tradeoff between these contradictory design threads. Our task in this section is precisely this: to explore the extension of a baseline 2D NoC implementation into the third dimension, while considering the aforementioned constraints.

9.2.1 A 3D Symmetric NoC Architecture

The natural and simplest extension to the baseline NoC router to facilitate a 3D layout is simply adding two additional physical ports to each router; one for Up and one for Down, along with the associated buffers, arbiters (VC arbiters and Switch Arbiters), and crossbar extension. We call this architecture a 3D Symmetric NoC, since both intra- and inter-layer movement bear identical characteristics: hop-by-hop traversal, as illustrated in Fig. 9.3a. For example, moving from the bottom layer of a 4-layer chip to the top layer requires three network hops.

This architecture, while simple to implement, has two major inherent drawbacks: (1) It wastes the beneficial attribute of a negligible inter-wafer distance (around 50 μm per layer) in 3D chips, as shown in Fig. 9.2. Since traveling in the vertical dimension is multi-hop, it takes as much time as moving within each layer. Of course, the average number of hops between a source and a destination does decrease as a result of folding a 2D design into multiple stacked layers, but inter-layer and intra-layer hops are indistinguishable. Furthermore, each flit must undergo buffering and arbitration at every hop, adding to the overall delay in moving up/down the layers. (2) The addition of two extra ports necessitates a larger

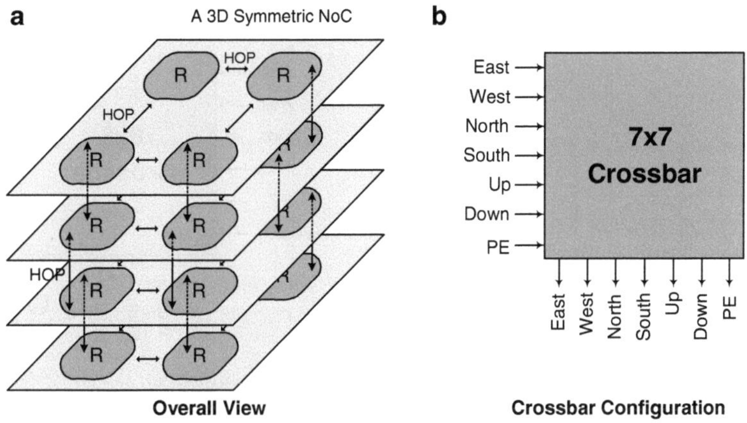

Fig. 9.3 A 3D symmetric NoC network. (**a**) Overall view. (**b**) Crossbar configuration

Table 9.1 Area and power comparisons of the crossbar switches assessed in Chapter 9

Crossbar type	Area	Power with 50% switching activity at 500 MHz
4×2 Crossbar(for 3D DimDe)	3039.32 μm²	1.63 mW
5×5 Crossbar (conventional 2D router)	8523.65 μm²	4.21 mW
6×6 Crossbar (3D NoC-bus hybrid)	11579.10 μm²	5.06 mW
7×7 Crossbar (3D symmetric NoC router)	17289.22 μm²	9.41 mW

7×7 crossbar, as shown in Fig. 9.3b. Crossbars scale upward very inefficiently, as illustrated in Table 9.1. This table includes the area and power budgets of all crossbar types investigated in this chapter, based on synthesized implementations in 90 nm technology. Details of the design and synthesis methodology are given in Section 9.4.2. Clearly, a 7×7 crossbar incurs significant area and power overhead over all other architectures. Therefore, the 3D Symmetric NoC implementation is a somewhat naive extension to the baseline 2D network.

9.2.2 *The 3D NoC-Bus Hybrid Architecture*

The previous sub-section argues that multi-hop communication in the vertical (inter-layer) dimension is not desirable. Given the very small inter-strata distance, single-hop communication is, in fact, feasible. This realization opens the door to a very popular shared-medium interconnect, the bus. The NoC router can be hybridized with a bus link in the vertical dimension to create a 3D NoC-Bus Hybrid structure, as shown in Fig. 9.4a. This approach was first introduced by the author in [43] and presented in Chapter 8 of this volume, where it was used in a 3D NUCA L2 Cache for CMPs. This hybrid system provides both performance and area benefits. Instead of an unwieldy 7×7 crossbar, it requires a 6×6 crossbar (Fig. 9.4b), since

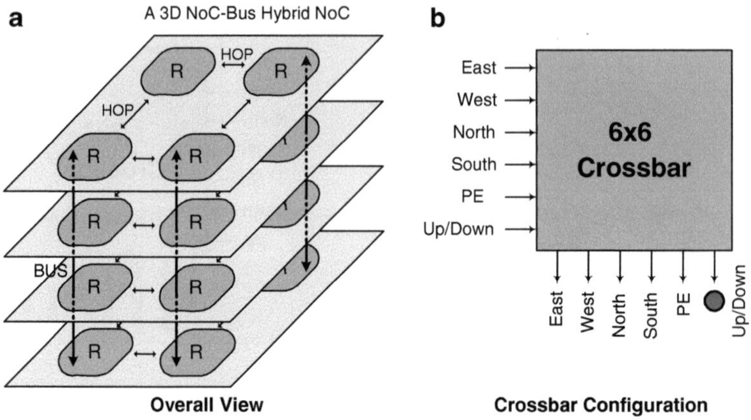

Fig. 9.4 A 3D NoC-bus hybrid architecture. (**a**) Overall view. (**b**) crossbar configuration

Fig. 9.5 Side view of the inter-layer via structure in a 3D NoC-bus hybrid structure

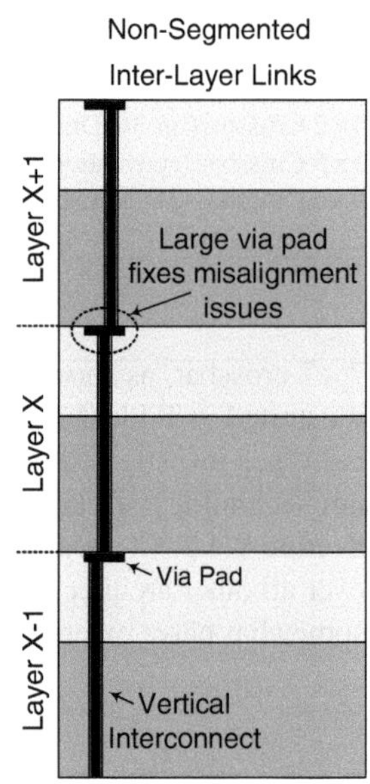

the bus adds a single additional port to the generic 2D 5 × 5 crossbar. The additional link forms the interface between the NoC domain and the bus (vertical) domain. The bus link has its own dedicated queue, which is controlled by a central arbiter. Flits from different layers wishing to move up/down should arbitrate for access to the shared medium.

Figure 9.5 illustrates the side view of the vertical via structure. This schematic depicts the usefulness of the large via pads between the different layers; they are deliberately oversized to cope with misalignment issues during the fabrication process. Consequently, it is the large vias that ultimately limit vertical via density in 3D chips.

Despite the marked benefits over the 3D Symmetric NoC router of Section 9.2.1, the bus approach also suffers from a major drawback: it does not allow concurrent communication in the third dimension. Since the bus is a shared medium, it can only be used by a single flit at any given time. This severely increases contention and blocking probability under high network load, as will be demonstrated in Section 9.4. Therefore, while single-hop vertical communication does improve performance in terms of overall latency, inter-layer bandwidth suffers.

9.2.3 A True 3D NoC Router

Moving beyond the previous options, we can envision a true 3D crossbar implementation, which enables seamless integration of the vertical links in the overall router operation. Figure 9.6 illustrates such a 3D crossbar layout. It should be noted at this point that the traditional definition of a *crossbar* – in the context of a 2D physical layout – is a switch in which each input is connected to each output through a single connection point. However, extending this definition to a physical 3D structure would imply a switch of enormous complexity and size (given the increased numbers of input- and output-port pairs associated with the various layers). Therefore, in this chapter, we chose a simpler structure which can accommodate the interconnection of an input to an output port through more than one connection points. While such a configuration can be viewed as a multi-stage switching network, we still call this structure a *crossbar* for the sake of simplicity.

The vertical links are now embedded in the crossbar and extend to all layers. This implies the use of a 5×5 crossbar, since no additional physical channels need to be dedicated for inter-layer communication. As shown in Table 9.1, a 5×5 crossbar

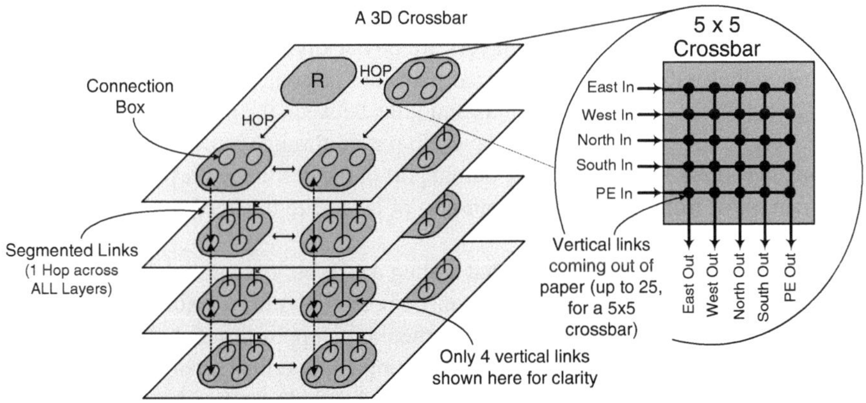

Fig. 9.6 NoC routers with true 3D crossbars

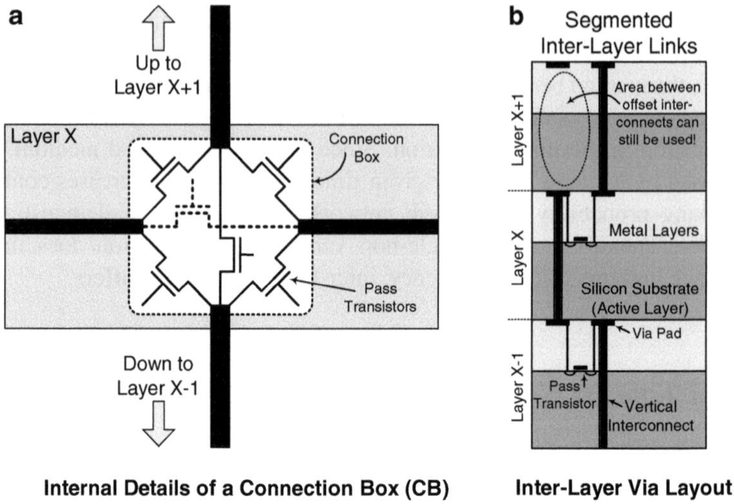

Fig. 9.7 Side view of the inter-layer via structure in a 3D crossbar. (**a**) Internal details of a connection box (CB). (**b**) Inter-layer via layout

is significantly smaller and less power-hungry than the 6×6 crossbar of the 3D NoC-Bus Hybrid and the 7×7 crossbar of the 3D Symmetric NoC. Interconnection between the various links in a 3D crossbar would have to be provided by dedicated connection boxes at each layer. These connecting points can facilitate linkage between vertical and horizontal channels, allowing flexible flit traversal within the 3D crossbar. The internal configuration of such a Connection Box (CB) is shown in Fig. 9.7a. The horizontal pass transistor is dotted, because it is not needed in our proposed 3D crossbar implementation, which is presented in Section 9.3. The vertical link segmentation also affects the via layout, as illustrated in Fig. 9.7b. While this layout is more complex than that shown in Fig. 9.5, the area between the offset vertical vias can still be utilized by other circuitry, as shown by the dotted ellipse in Fig. 9.7b.

Hence, the 2D crossbars of all layers are physically fused into one single three-dimensional crossbar. Multiple internal paths are present, and a traveling flit goes through a number of switching points and links between the input and output ports. Moreover, flits re-entering another layer do not go through an intermediate buffer; instead, they directly connect to the output port of the destination layer. For example, a flit can move from the western input port of layer 2 to the northern output port of layer 4 in a single hop.

It will be shown in Section 9.3 that adding a 128-bit vertical link, along with its associated control signals, consumes only about 0.01 mm^2 of silicon real estate. However, despite this encouraging result, there is an opposite side to the coin which paints a rather bleak picture. Adding a large number of vertical links in a 3D crossbar to increase NoC connectivity results in increased path diversity. This translates into multiple possible paths between source and destination pairs. While this increased diversity may initially look like a positive attribute, it actually leads to a dramatic increase in the complexity of the central arbiter, which coordinates inter-layer communication in the 3D crossbar. The arbiter now needs to decide between a multitude of possible interconnections, and requires an excessive number of control signals to enable all these interconnections. Even if the arbiter functionality can be distributed to multiple smaller arbiters, then the coordination between these arbiters becomes complex and time-consuming. Alternatively, if dynamism is sacrificed in favor of static path assignments, the exploration space is still daunting in deciding how to efficiently assign those paths to each source-destination pair. Furthermore, a full 3D crossbar implies 25 (i.e., 5×5) Connection Boxes (see Fig. 9.7a) per layer. A four-layer design would, therefore, require 100 CBs! Given that each CB consists of six transistors, the whole crossbar structure would need 600 control signals for the pass transistors alone! Such control and wiring complexity would most certainly dominate the whole operation of the NoC router. Pre-programming static control sequences for all possible input-output combinations would result in an oversize table/index; searching through such a table would incur significant delays, as well as area and power overhead. The vast number of possible connections hinders the otherwise streamlined functionality of the switch. Note that the prevailing tendency in NoC router design is to minimize operational complexity in order to facilitate very short pipeline lengths and very high frequency. A full crossbar with its overwhelming control and coordination complexity poses a stark contrast to this frugal and highly efficient design methodology. Moreover, *our experimental results will show that the redundancy offered by the full connectivity is rarely utilized by real-world workloads, and is, in fact, design overkill.*

To understand the magnitude of the path diversity issue in a true 3D crossbar (as shown in Fig. 9.8 for a $3 \times 3 \times 3$ example), one can picture the 3D crossbar itself as a 3D Mesh network. For the 3D $3 \times 3 \times 3$ crossbar of Fig. 9.8, the number of minimal paths, k, between points A and B is given in [82] as

$$k = \binom{\Delta_x + \Delta_y + \Delta_z}{\Delta_x}\binom{\Delta_y + \Delta_z}{\Delta_y} = \frac{(\Delta_x + \Delta_y + \Delta_z)!}{\Delta_x!\Delta_y!\Delta_z!} \tag{9.1}$$

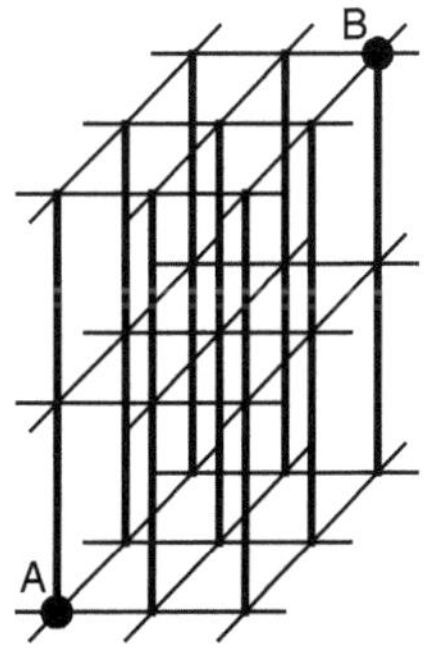

Fig. 9.8 A 3D $3 \times 3 \times 3$ crossbar in conceptual form

where Δx, Δy and Δz are the numbers of hops separating A and B in the X, Y, and Z dimensions, respectively. In our example, $\Delta x = \Delta y = \Delta z = 2$. Thus, the number of minimal paths between A and B is 90. For a 3D $4 \times 4 \times 4$ crossbar, this number explodes to 1,680. If non-minimal paths are also considered, then path diversity is practically unbounded [82].

Hence, given the tight latency and area constraints in NoC routers, vertical (inter-layer) arbitration should be kept as simple as possible. This can be achieved by using a limited amount of inter-layer links. The question is then: *how many links are enough? Our experiments in Section 9.4 demonstrate that anything beyond two links per 3D crossbar yields diminishing returns in terms of performance.*

9.2.4 A Partially-Connected 3D NoC Router Architecture

The scalability problem in vertical link arbitration highlighted in the previous subsection dictates the use of a partially-connected 3D crossbar, i.e., a crossbar with a limited number of vertical links. The arbitration complexity can be further mitigated through the use of hierarchical arbiters. Two types of arbiters should be employed: intra-layer arbiters, which handle local requests from a single layer, and one global arbiter per vertical link to handle requests from all layers. This decoupling of arbitration policies can help parallelize tasks; while flits arbitrate locally in each layer, vertical arbitration decides on inter-layer traversal. These design directives were the fundamental drivers in our quest for a suitable 3D NoC implementation. As such, they form the cornerstones of our proposed architecture, which is described in detail in the following section.

9.3 The Proposed 3D Dimensionally-Decomposed (DimDe) NoC Router Architecture

The heart of a typical two-dimensional NoC router is a monolithic, 5×5 crossbar, as depicted abstractly in Fig. 9.9a. The five inputs/outputs correspond to the four cardinal directions and the connection from the local PE. The realization that the crossbar is a major contributor to the latency and area budgets of a router has fueled extensive research in optimized switch designs. Through the use of the preliminary switching process described in Chapter 4 – known as Guided Flit Queuing [40] – incoming traffic may be decomposed into two independent streams: (a) East–West traffic (i.e., packet movement in the X dimension), and (b) North–South traffic (i.e., packet movement in the Y dimension). This segregation of traffic flow allows the use of two smaller 2×2 crossbars and the isolation of the two flows in two independent router sub-modules, as shown conceptually in Fig. 9.9b. The resulting two compact modules are more area- and power-efficient, and provide better performance than the conventional monolithic approach.

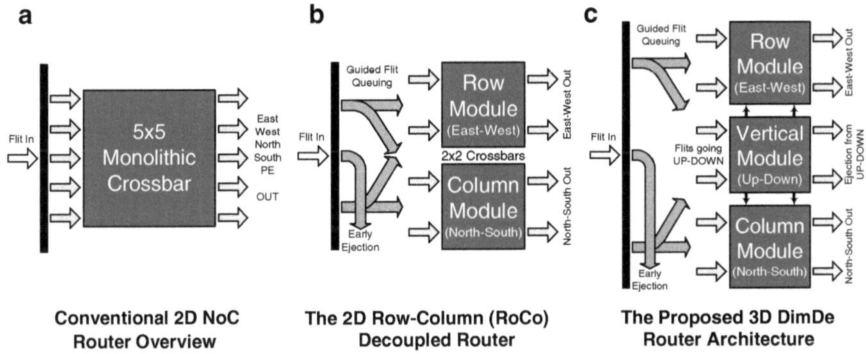

Fig. 9.9 Different NoC router switching mechanisms. (**a**) Conventional 2D NoC router overview. (**b**) The 2D Row–Column (RoCo) decoupled router. (**c**) The proposed 3D DimDe router architecture

Following this logic of traffic decomposition in orthogonal dimensions, we propose in this work the addition of a third information flow in the Z dimension (i.e., inter-layer communication). An additional module is now required to handle all traffic in the third dimension; this component is aptly called the Vertical Module. On the input side, packets are decomposed into the three dimensions (X, Y, and Z), and forwarded to the appropriate module. However, as previously mentioned, simply adding a third independent module cannot lead to a true 3D crossbar, because inter-layer traffic must be able to merge with intra-layer traffic upon arrival at the destination chip layer. A totally decoupled Vertical Module would force all packets arriving at a particular layer and wishing to continue traversal within that layer to be re-buffered and re-arbitrate for access to the Row/Column modules.

Hence, the Vertical Module must somehow fuse the Row and Column modules to allow movement of packets from the Vertical Module to the Row and Column Modules. An abstract view of the proposed 3D DimDe implementation is illustrated in Fig. 9.9c. The diagram clearly shows the Vertical Module linking with the Row and Column Modules. Also notice that the communication link is one-way, i.e., from the Vertical Module to the Row/Column Modules. There is no need for the Row/Column Modules to communicate with the Vertical Module, since intra-layer traffic wishing to change layer is pre-directed to the Vertical Module at the input of the router.

The streamlined nature of a dimensionally decomposed router lends itself perfectly for a 3D crossbar implementation. The simplicity and compactness of the smaller, distinct modules can be utilized to create a crossbar structure which extends into the third dimension without incurring prohibitive area and latency overhead. The high-level architectural overview of our proposed 3D DimDe router is shown in Fig. 9.10. As illustrated in the figure, the gateway to different layers is facilitated by the inclusion of the third, Vertical Module. The 3D DimDe router uses vertical links which are segmented at the different device layers through the use of compact Connection Boxes (CB). Figure 9.7a shows a side view cross-section of such a CB. Each box consists of five pass transistors which can connect the vertical

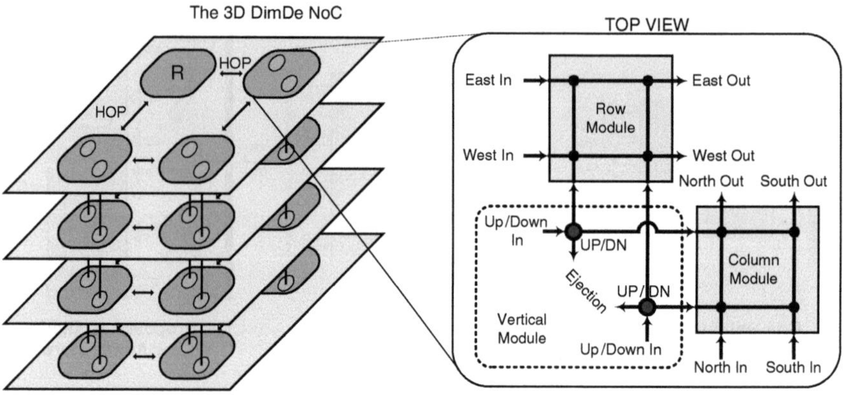

Fig. 9.10 Overview of the 3D DimDe NoC architecture

Table 9.2 The effect of inter-layer distance on propagation delay

Inter-layer link length	Number of repeaters	Delay
50 μm (layer 1 to 2)	0	7.86 ps
100 μm (layer 1 to 3)	0	19.05 ps
150 μm (layer 1 to 4)	0	36.12 ps
150 μm (layer 1 to 4)	1 (layer 3)	105.14 ps

(inter-layer) links to the horizontal (intralayer) links. The dotted transistor is not needed in DimDe, because the design was architected in such a way as to avoid the case where intra-layer communication needs to pass through a CB. The CB structure allows simultaneous transmission in two directions, e.g., a flit coming from layer X + 1 and connecting to the left link of Layer X, and a flit coming from layer X-1 connecting to the right link of layer X (see Fig. 9.7a). The inclusion of pass transistors in the data path adds delay and degrades the signal strength due to the associated voltage drop. However, this design decision is fully justified by the fact that inter-layer distances are, in fact, negligible. To investigate the effectiveness and integrity of this connection scheme, we laid out the physical design of the CB and simulated it in HSpice using the Predictive Technology Model (PTM) [146] at 70 nm technology and 1 V power supply. The latency results for 2, 3 and 4-layer distances are shown in Table 9.2. Evidently, even with a four-layer design (i.e., traversing four cascaded pass transistors), the delay is only 36.12 ps; this is a mere 1.8% of the 2 ns clock period (500 MHz) of the NoC router. In fact, the addition of repeaters will increase latency, because with such small wire lengths (around 50 μm per layer), the overall propagation delay is dominated by the gate delays and not the wiring delay. This effect is corroborated by the increased delay of 105.14 ps – when using a single repeater – in Table 9.2.

To indicate the fact that each vertical link in the proposed architecture is composed of a number of wires, we thereby refer to these links as bundles. The presence of a segmented wire bundle dictates the use of one central arbiter for each vertical bundle, which is assigned the task of controlling all traffic along the vertical link.

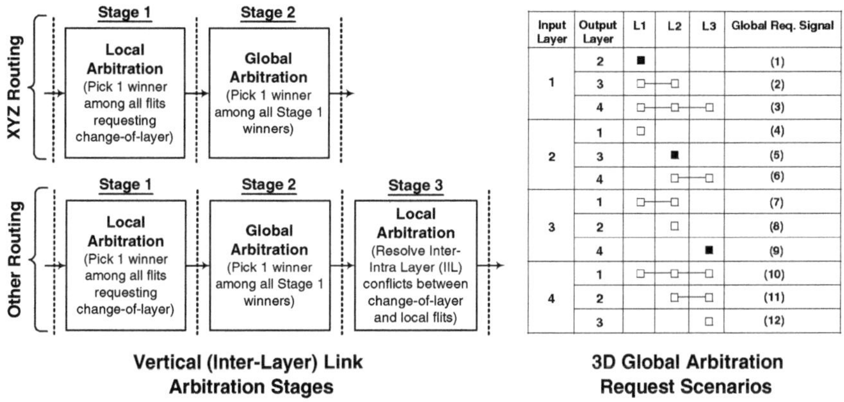

Fig. 9.11 Vertical (inter-layer) link arbitration details. (**a**) Vertical (inter-layer) link arbitration stages. (**b**) 3D global arbitration request scenarios

If arbitration is carried out at a local level alone, then the benefit of concurrent communication along a single vertical bundle cannot be realized; each layer would simply be unaware of the connection requests of the other layers. Hence, a coordinating entity is required to monitor all requests for vertical transfer from all the layers and make an informed decision, which will favor simultaneous data transfer whenever possible. Concurrent communication increases the vertical bandwidth of the 3D chip. Given the resource-constrained nature of NoCs, however, the size and operational complexity of the central arbiter should be handled judiciously. The goal is not to create an overly elaborate mechanism which provides the best possible matches over several clock cycles. Our objective was to obtain reasonably intelligent matches within a single clock cycle.

To achieve this objective, we divided the arbitration for the vertical link into two stages, as shown at the top of Fig. 9.11a. The first stage is performed locally, within each layer. This stage arbitrates over all flits in a single layer which request a transfer to a different layer. Once a local winner is chosen, the local arbiter notifies the second stage of arbitration, which is performed globally. This global stage takes in all winning requests from each layer and decides on how the segmented link will be configured to accommodate the inter-layer transfer(s). The arbiter was designed in such a way as to realize the scenarios which are suitable for concurrent communication.

Figure 9.11b illustrates all possible requests to the global arbiter of a particular vertical bundle, assuming a 4-layer chip configuration using the deterministic XYZ routing. The designations L1, L2, and L3 indicate the different segments of the vertical bundle; L1 is the link between layers 1 and 2, L2 is the link between layers 2 and 3, and so on. As an example, let us assume that a flit in layer 1, which wants to go to layer 2, has won the local arbitration of layer 1; global request signal 1 (see Fig. 9.11b) is asserted. Similarly, a flit in layer 2 wants to go to layer 3; global request signal 5 is asserted. Finally a flit in layer 3 wants to go to layer 4; global request signal 9 is asserted. The global arbiter is designed to recognize that the global request combination 1, 5, 9 (black boxes in Fig. 9.11b) results in full concurrent communication between all participating layers. It will, therefore, grant all requests simultaneously. All combinations which favor simultaneous, non-overlapping communication are programmed into the global arbiter. If needed, these configurations can be given higher priority in the selection process. The arbiter can be placed on any layer, since the vertical distance to be traveled by the inter-layer control signals is negligible.

The aforementioned two arbitration stages suffice only if deterministic XYZ routing is used. In this case, a flit traveling in the vertical (i.e., Z) dimension will be ejected to the local PE upon arrival at the destination layer's router. If, however, a different routing algorithm is used, which allows flits coming from different layers to continue their traversal in the destination layer, then an additional local arbitration stage is required to handle conflicts between flits arriving from different layers and flits residing in the destination layer. The third arbitration stage, illustrated at the bottom of Fig. 9.11a, will take care of such Inter-Intra Layer (IIL) conflicts. The use of non-XYZ algorithms also complicates the request signals sent across different layers. It is no longer enough to merely indicate the destination layer; the output port designation on the destination layer also needs to be sent. IIL conflicts highlight the complexity involved in coordinating flit traversal in a 3D network environment. An example of the use of a non-XYZ routing algorithm is presented in Fig. 9.12, which tracks the path of a flit traveling from Layer X to the eastern output of Layer X+1. In this case, the flit changes layer and continues traversal in a different layer.

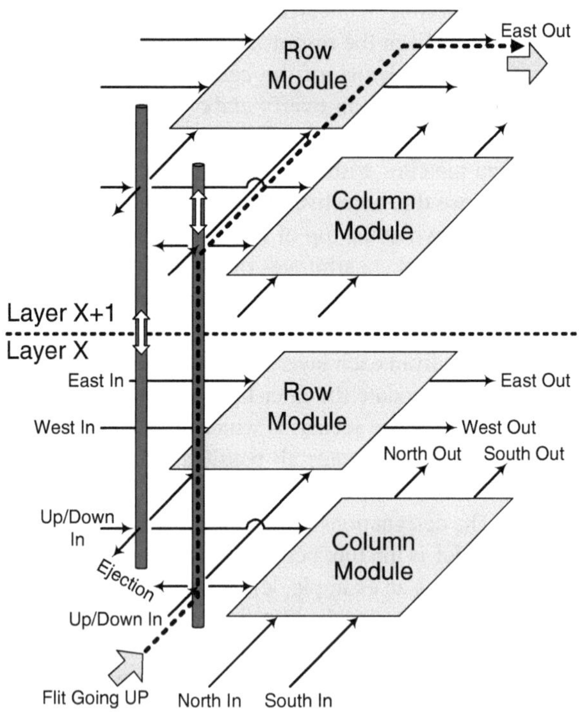

Fig. 9.12 An example of a non-XYZ routing algorithm

Each vertical bundle in DimDe consists of a number of data wires (128 bits in this work), and a number of control wires to/from a central arbiter, which coordinates flit movement in the vertical dimension. These control signals include: (a) Request signals from all layers to the central arbiter indicating the requested destination layer (and possibly output port, depending on the routing algorithm used), and the corresponding acknowledgement signals from the arbiter. (b) Enable signals from the arbiter to the pass transistors of the Connection Boxes of each layer spanned by the wire bundle. The total number of wires, w, in a vertical bundle is given by

$$w = \begin{cases} b + 2(n-1)^2 + 5(n-1), & \textit{if XYZ algorithm} \\ b + 2(n-1)^2 + 6(n-1) + 5(n-1), & \textit{otherwise} \end{cases} \tag{9.2}$$

where

b = *number of data bits/wires,*

$2(n-1)^2$ = *number of request/acknowledgement signals to/from the central arbiter assuming an n-layer chip,*

$6(n-1)$ = *number of additional signals sent to/from the arbiter for output port designation* (3-*bit designation for the four possible output ports and the ejection port) when a non-XYZ routing algorithm is employed,*

$5(n-1)$ = *number of enable signals for the pass transistors of the CB of each layer.*

Assuming a 4-layer configuration (n=4), XYZ routing, and 128 data bits (i.e., b=128), the number of wires in a vertical bundle, w, is 161. Based on the square-like layout of Fig. 9.2, the area consumed by the bundle is around 10,000 μm^2=0.01 mm^2. This amounts to a vertical via density of around 1.5 million individual wires per cm^2. This result illustrates the fact that increasing the number of vertical vias is, in fact, feasible in terms of area consumption by the wires themselves. However, as explained in Section 9.2.3, adding extra vertical bundles in the 3D crossbar is prohibitive in terms of arbitration complexity; the area, power and latency increases incurred by a highly-complex arbitration scheme negate any advantages provided by the increased number of interlayer bundles. Furthermore, it will be demonstrated later on that increasing the number of inter-layer bundles yields rapidly diminishing returns in terms of performance gain under both synthetic and real workloads.

A detailed view of the proposed 3D DimDe architecture is shown in Fig. 9.13. DimDe employs Guided Flit Queuing to guide incoming flits to an appropriate Path Set (PS). Guided Flit Queuing is a preliminary switching operation at the input of the router which utilizes the look-ahead routing information present in incoming header flits. This information denotes the requested output path; thus, incoming traffic can be decomposed into the X, Y, and Z dimensions. The Vertical Module adds two extra path sets to the 2D implementation. One path set is used by incoming flits from the East–West (intra-layer) dimension, and the other for flits from the North–South dimension. Just like Guided Flit Queuing, the Early Ejection Mechanism (see Chapter 4 and [40]) uses the look-ahead routing information to identify packets which need to be ejected to the local PE. This enables

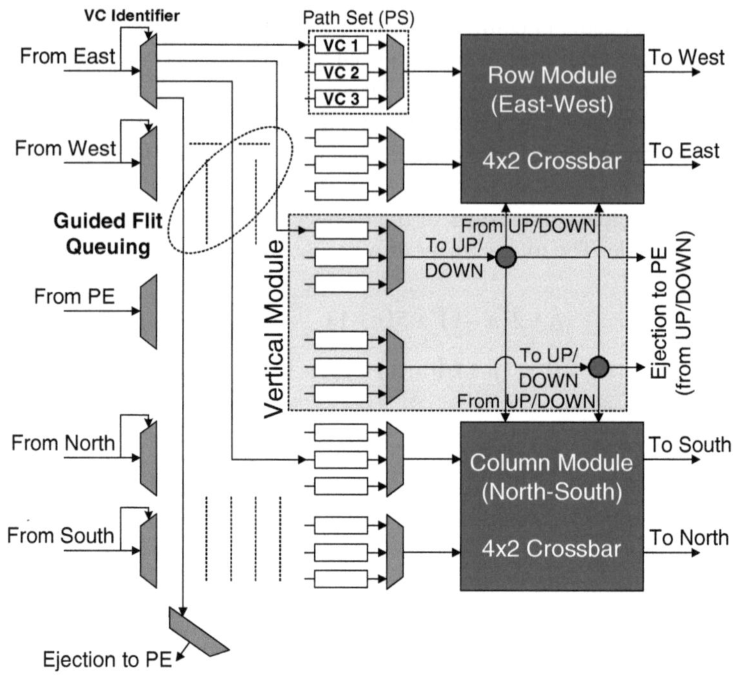

Fig. 9.13 Architectural detail of the proposed 3D DimDe NoC router

such flits to bypass the destination router and be directly ejected to the NIC. The Vertical Module consists of two bidirectional vertical bundles, one for each of the two path sets. Note that the number of vertical bundles can be varied from four to one. Each vertical link has one input connection and three output connections on each layer. The input connection comes from the associated path set's MUX (see dark box in the middle of Fig. 9.13). The three output connections are as follows: (1) One connection to the Row Module Crossbar for flits which arrive from other layers and need to continue traversal in the East–West dimension of the current layer. (2) One connection to the Column Module crossbar for flits which need to continue in the North–South dimension of the current layer. (3) One connection for ejection to the Network Interface Controller (NIC) of the local PE. This configuration implies that the Row and Module crossbars need to grow in size from 2×2 in the 2D case to 4×2 in DimDe to accommodate the two additional connections from the two vertical links. Despite this increase in size, two 4×2 crossbars are still substantially smaller than a single monolithic 6×6 or 7×7 crossbar, as illustrated in Table 9.1. Once again, it is precisely for this reason that we chose to use this architecture in our 3D NoC implementation.

The Vertical Module of the proposed DimDe router uses two Path Sets to group the available Virtual Channels. As shown in Fig. 9.14, the DimDe router requires 5 VCs for correct functionality under a deterministic, deadlock-free algorithm: one VC for injection from each of the four incoming directions, and one for injection from the local PE. The sixth VC can be used as a drain channel for deadlock recovery

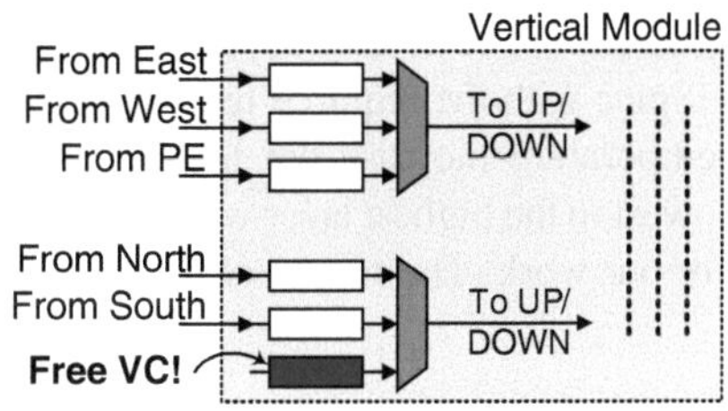

Fig. 9.14 Virtual channel assignments in the vertical module of DimDe

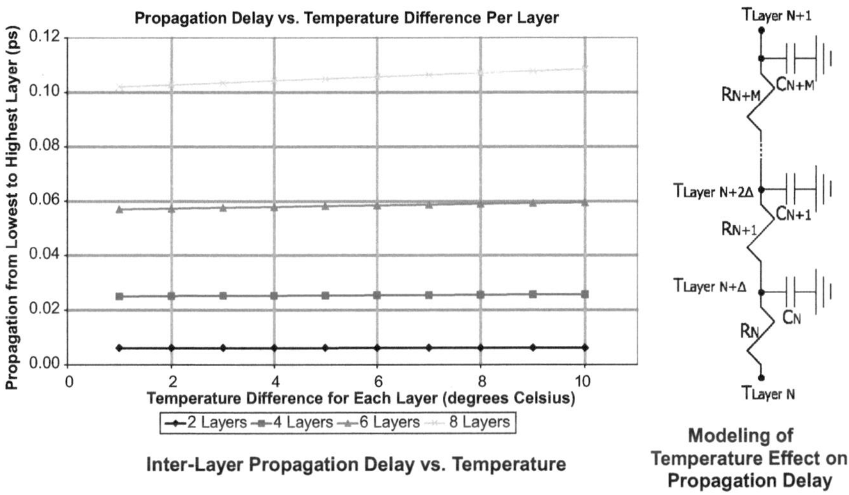

Fig. 9.15 Thermal effects on inter-layer propagation delay. (**a**) Inter-layer propagation delay vs. temperature. (**b**) Modeling of temperature effect on propagation delay

under adaptive routing algorithms. Moreover, depending on the algorithm used, additional VCs can be added to the two Vertical Module path sets to ensure deadlock freedom. These drain VCs need to operate on deadlock-free algorithms to guarantee deadlock breakup [71]. In this work, we concentrated on deterministic XYZ and ZXY algorithms as a proof of concept of the proposed architecture. Since these algorithms are inherently deadlock-free, the sixth VC buffer was used as an additional injection VC from the local PE.

As previously explained in Chapter 8, thermal issues are of utmost importance in 3D chips. Stacking several active layers with minimal distance in-between favors the creation of hotspots. From a 3D NoC perspective, it was important to investigate the effect of high temperature on the propagation delay of the signals on the vertical (inter-layer) interconnects. To that extent, the propagation delay between the layers was modeled as an RC ladder (Fig. 9.15b) to accurately capture the distributed resistance, capacitance, and temperature variations along the inter-strata vias. The resistance of metals is affected by temperature, and it was modeled using equations from [184]. Assuming a T_{Layer1} temperature of 85°C and a fixed linear temperature gradient between each layer, the propagation delay of these vias was simulated in HSpice with the required temperature annotations. Even in the worst case of a 10°C temperature increase per layer for 8 layers, the total propagation delay from the lowest to the highest layer was only 0.11 ps and, therefore, considered inconsequential for our work. The results of the thermal analyses are summarized in Fig. 9.15a.

9.4 Performance Evaluation

This section presents simulation-based performance evaluation of the proposed architecture, a generic 2D router architecture, a 3D Symmetric NoC design, the 3D NoC-Bus Hybrid architecture, and the Full 3D Crossbar implementation, in terms of network latency, throughput and power consumption under various traffic patterns. The experimental methodology is followed by the associated results.

9.4.1 Simulation Platform

A double-faceted evaluation environment was implemented in order to conduct a detailed evaluation of the router architectures analyzed in this chapter:

(a) A cycle-accurate stand-alone 3D NoC simulator was developed, which accurately models the routers, the interconnection links and vertical pillars, as well as all the architectural features of the various NoC architectures under investigation. The simulator was built by augmenting an existing 2D NoC simulator and models each individual component within the router architecture, allowing for detailed analysis of component utilizations and flit flow through the network. The activity factor of each component is used for analyzing power consumption within the network. In addition to the network-specific parameters, our simulator accepts hardware parameters such as power consumption (dynamic and leakage) for each component and overall clock frequency. This leg of the simulation process examines the behavior of all the architectures under synthetic workloads.

(b) To provide a more diversified simulation environment, we also implemented a detailed trace-driven cycle-accurate hybrid NoC/cache simulator for CMP architectures. The memory hierarchy implemented is governed by a two-level directory cache coherence protocol. Each core has a private write-back L1 cache (split L1 I and D cache, 64 KB, 2-way, 3-cycle access). The L2 cache is shared among all cores and split into banks (32 banks, 512 KB each for a total of 16 MB, 6-cycle bank access). An underlying NoC model connects the L2 banks. The L1/L2 block size is 64 B. Our coherence model includes a MESI-based protocol with distributed directories, with each L2 bank maintaining its own local directory. The simulated memory hierarchy mimics SNUCA [158]. The sets are statically placed in the banks depending on the low order bits of the address tags. The network timing model simulates all kinds of messages: invalidates, requests, replies, write-backs, and acknowledgements. The interconnect model is the same as (a) above. The off-chip memory is a 4 GB DRAM with a 260-cycle access time.

Detailed instruction traces of four commercial server workloads were used: (1) *TPC-C* [185], a database benchmark for online transaction processing (OLTP), (2) *SAP* [186], a sales and distribution benchmark, and (3) *SJBB* [187] and (4) *SJAS* [188], two Java-based server benchmarks. The traces – collected from multiprocessor server configurations at Intel Corporation – were then run through our NoC/cache hybrid simulator to measure network statistics. Additionally, a second set of memory traces was generated by executing programs from *SPLASH* [189], a suite of parallel scientific benchmarks, on the Simics full system simulator [153]. Specifically, *barnes*, *ocean*, *water-nsquared* (*wns*), *water-spatial* (*wsp*), *lu*, and *radiosity* (*rad*) were used. The baseline configuration is the Solaris 9 Operating system running on eight UltraSPARC III cores. Benchmarks execute 16 parallel threads. Again, the number of banks for the L2 shared cache is 32. Thus, 32 nodes are present in the NoC network, 8 of which are also CPU nodes.

9.4.2 Energy Model

The proposed components of the 3D router architectures, and a generic two-stage 5-port router architecture were implemented in structural Register-Transfer Level (RTL) Verilog and then synthesized in Synopsys Design Compiler using a TSMC 90 nm standard cell library. The library utilized appropriate wire-load approximation models to reasonably capture wire loading effects. The vertical interconnects were modeled as 2D wires with equivalent resistance and capacitance. The resulting designs operate at a supply voltage of 1 V and a clock speed of 500 MHz. Both dynamic and leakage power estimates were extracted from the synthesized router implementation. These power numbers were then imported into our cycle-accurate simulation environment and used to trace the power profile of the entire on-chip network.

9.4.3 Performance Results

The proposed 3D DimDe design was compared against four other router architectures (2D NoC, 3D Symmetric NoC, 3D NoC-Bus Hybrid, and a Full 3D Crossbar configuration) using our cycle-accurate simulation environment. Our definition of a "full" 3D crossbar implies that all connection points inside the 2D 5 × 5 crossbar (i.e., 25 links) extend into the third (i.e., vertical) dimension. It should *not* be confused with a *non-blocking* crossbar where each input can be connected to each output regardless of how the other inputs and outputs are interconnected. Such a configuration would be tremendously complex in a physical 3D setting because the number of possible input-output pairs explodes as the number of layers increases.

In both simulation phases, two deterministic routing algorithms (XYZ routing and ZXY routing) were used to measure the average network latency, throughput, and power consumption in all experiments. For the synthetic workload simulation phase (described in part (a) of Section 9.4.1), all architectures under investigation were evaluated using a regular mesh network with 64 nodes. In the 3D designs, 4 layers were used, each with 16 nodes (4×4). Wormhole routing [55] based on virtual-channel flow control [57,190] was employed in all cases. To ensure fairness, all architectures under test had 3 VCs per input port, and a total buffer space of 80 flits per node. Each simulation consists of two phases: a warm-up phase of 20,000 packet injections, followed by the main phase which injects a further one million packets. Each packet consists of four 128-bit flits. The simulation terminates when all packets are received at the destination nodes. Uniform, matrix-transpose (dimension reversal) [191] and self-similar traffic patterns were used.

For the real workload simulation phase (described in part (b) of Section 9.4.1), the 32 L2 cache banks (nodes) were "folded" into 4 layers, with each layer holding 8 banks (4×2). The 8 CPUs were also split into 2 CPUs/layer. The commercial workloads were simulated for 10,000 transactions per thread, whereas the scientific workloads were simulated for 100 million instructions per core upon commencement of the parallel phase of the code. Data messages were 5-flit packets (64 B cache-line plus network overhead), while control messages were single-flit packets.

The 3D DimDe architecture design exploration provided different options for the number of pillars in the Vertical Module. Since we are using 2×2 crossbars as the basic building blocks, four pillars would provide a complete crossbar connection, while a single pillar would provide a segmented bus connection. As previously mentioned, the caveat is that more vertical pillars offer more path diversity and complicate the arbiter design. Hence, the number of pillars should be decided based on the performance, energy and area tradeoffs. Figure 9.16 illustrates the effect of the number of vertical pillars (per node) on average network latency. Interestingly,

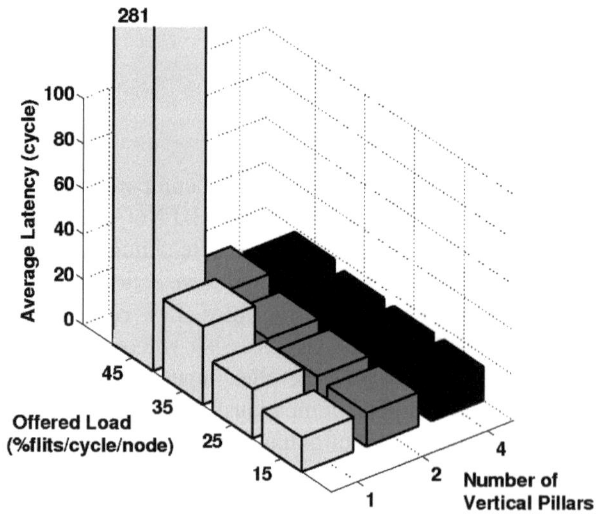

Fig. 9.16 Impact of the number of vertical bundles on performance

going from two to four vertical bundles yields rapidly diminishing returns in terms of performance gains. This experiment suggests that the two-pillar DimDe design provides the best compromise in terms of performance, area and energy behavior; employing only two – instead of more – vertical links would lower design complexity and power consumption without adversely affecting network latency. Therefore, in the rest of the evaluations, we use the two-pillar DimDe architecture (as shown in Fig. 9.13). Moreover, to validate our assertion that anything more than two vertical bundles would yield diminishing returns, we will compare our design to a full 3D crossbar configuration (Section 9.2.3) with 25 vertical bundles. In all experiments, *the Full 3D crossbar was assumed to complete its configuration in a single clock cycle. Therefore, all results for this Full 3D Crossbar design will be very optimistic. However, despite discounting the complexity of the control and arbitration logic of the Full 3D crossbar, the proposed DimDe router will still be able to achieve comparable performance.*

The latency and throughput results of all five architectures for various synthetic traffic patterns (i.e., phase 1 of our simulation experiments) are illustrated in Figs. 9.17 and 9.18. It can be observed that the proposed DimDe design consistently remains within 5% (on average) of the ideal Full 3D Crossbar's performance, while providing much lower latency and saturating at much higher workloads than the remaining architectures. Compared to the DimDe design, the 3D Symmetric topology suffers from the additional router delay at each inter-layer hop. At low loads (e.g., up to 20% for all traffic patterns with XYZ routing), the NoC-Bus Hybrid provides lower latency compared to the 3D Symmetric NoC as it benefits from the single hop vertical communication. As the load increases, the NoC-Bus Hybrid Architecture exhibits the worst latency and throughput for all traffic patterns (even worse than the 2D topology) as the bus saturates first with higher workload. Consequently, the 3D NoC-Bus Hybrid may be suitable only for 3D architectures where the traffic is mostly confined to the 2D strata and the load on the vertical links is sparse. Clearly, *the proposed 3D DimDe router outperforms the other three designs in all traffic patterns, and it achieves performance very close to that of a full crossbar, using only two (instead of 25) vertical bundles. This soundly resonates our assertions that a full 3D crossbar is design overkill in terms of performance enhancement.* The results with ZXY routing follow the same trends as with XYZ routing; thus, they are omitted for brevity.

In terms of the throughput behavior (Fig. 9.18), the DimDe architecture provides 18% average improvement over the other designs, while remaining within around 3% of the Full 3D Crossbar's throughput.

Figures 9.19 and 9.20 show the results of phase 2 of our simulation experiments, i.e., real commercial and scientific workloads in an 8-CPU CMP environment. Figure 9.19 depicts the average network latency for the four 3D architectures under test. We do not show the 2D results, since the significantly larger hop count in the 2D case naturally leads to substantially worse results compared to all 3D architectures. Clearly, the proposed 3D DimDe design outperforms all designs except the Full 3D Crossbar. DimDe provides an average improvement of 27% over the 3D Symmetric and 3D NoC-Bus Hybrid designs, and remains within 4% of the Full 3D

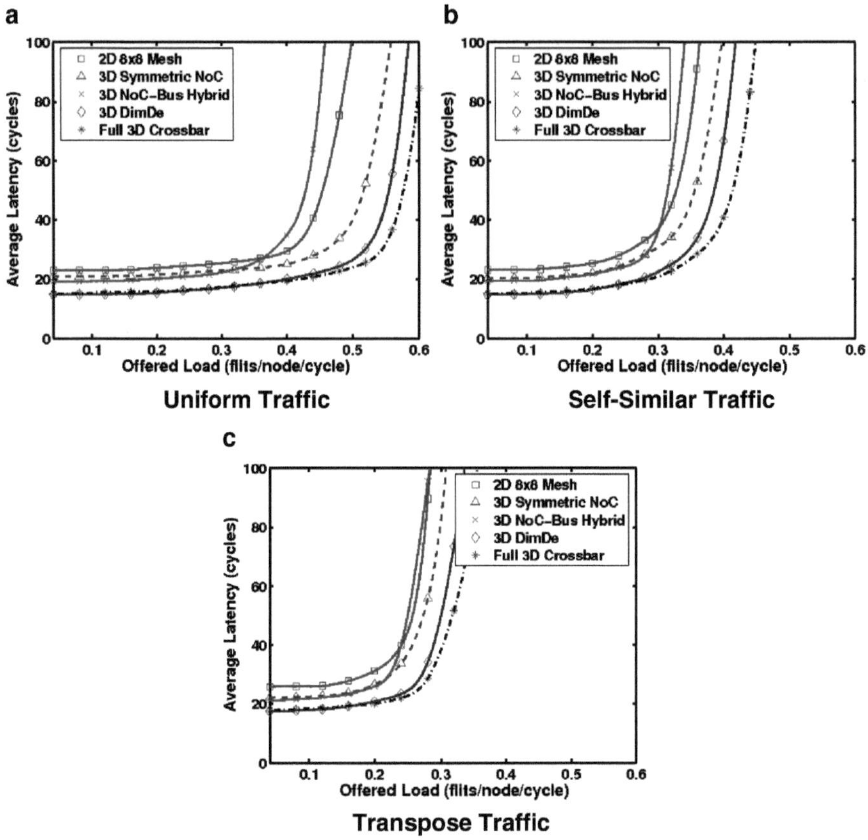

Fig. 9.17 Average latency with various synthetic traffic patterns (XYZ routing). (**a**) Uniform traffic. (**b**) Self-similar traffic. (**c**) Transpose traffic

Crossbar's performance. However, a more complete picture is painted in Fig. 9.20, which compares the Energy-Delay Product (EDP) of all the architectures. This metric is, in fact, more meaningful since it accounts for both performance and power consumption. Here, the efficiency of the proposed 3D DimDe design shines through. DimDe significantly outperforms all other designs, including the ideal Full 3D Crossbar, by 26% on average. These results underscore the efficiency of the DimDe architecture. *Through the decomposition of incoming traffic into smaller components, the use of a simple, partially-connected 3D crossbar, and reduced arbitration complexity, DimDe can outperform even the optimistic results of a full (i.e., 25 vertical bundles) crossbar structure. This result is of profound significance, because it shows that increasing inter-layer links arbitrarily increases design complexity and overhead without tangible performance benefits.*

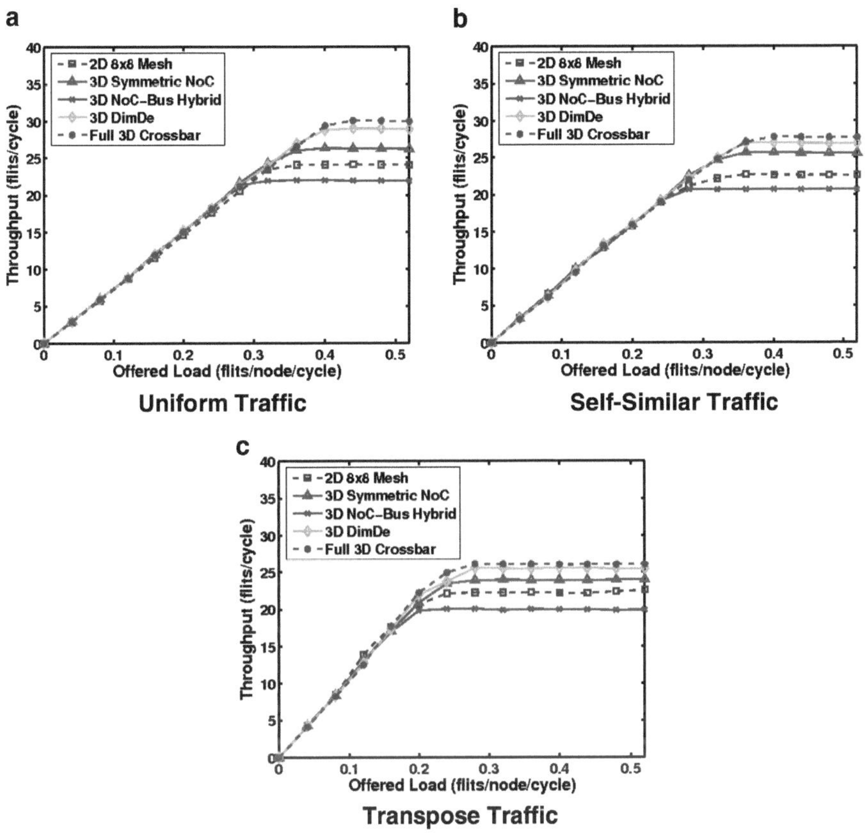

Fig. 9.18 Throughput with various synthetic traffic patterns (XYZ routing). (**a**) Uniform traffic. (**b**) Self-similar traffic. (**c**) Transpose traffic

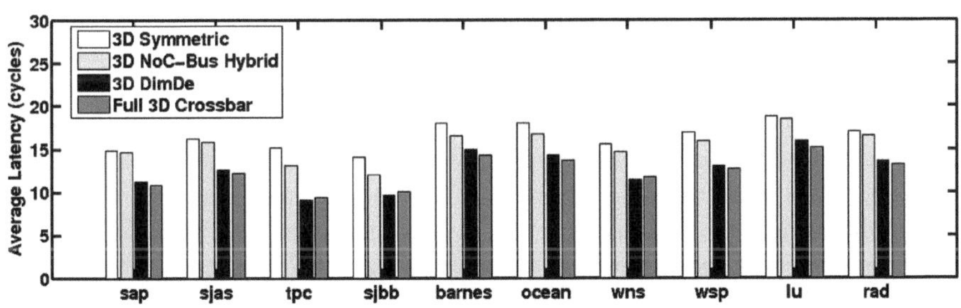

Fig. 9.19 Average latency with various commercial and scientific workloads

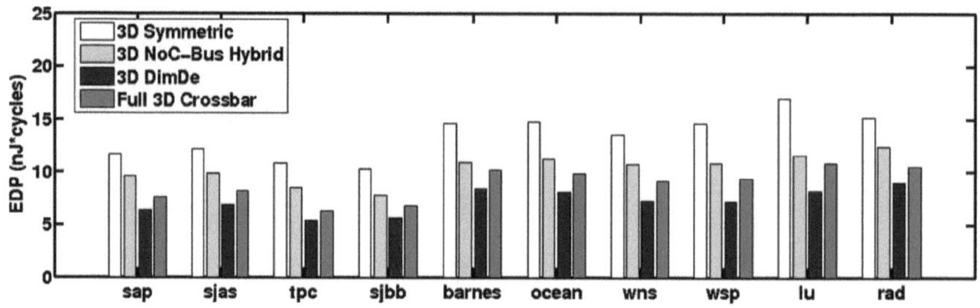

Fig. 9.20 Energy-delay product (EDP) with various commercial and scientific workloads

9.5 Chapter Summary

3D technology is envisioned to provide a performance-rich, area- and energy-efficient, and temperature-aware design space for multicore/SoC architectures. In this context, the on-chip interconnect in a 3D setting will play a crucial role in optimizing the performance, area, energy and thermal behaviors. In this chapter, several design options for 3D NoCs have been explored, specifically focusing on the inter-strata communication. Three possible designs that include a simple bus for the vertical connection, a symmetric 3D hop-by-hop topology, and a true 3D crossbar architecture are investigated. The proposed 3D architecture, called the 3D DimDe router, supports two vertical interconnects to achieve a balance between the path diversity and high bandwidth offered by a full 3D crossbar and the simplicity of a bus. DimDe supports a true 3D crossbar structure spanning all layers of the chip and fusing them into a single router entity. We have investigated the detailed micro-architectural implications of the design, which include the feasibility of the inter-strata vertical wire layout, arbitration mechanism, and virtual channel support for providing deadlock-free routing. The design has been implemented in structural Verilog and synthesized using a TSMC 90 nm standard cell library to analyze the area, energy, and thermal behaviors. It has been shown that it is possible to implement a hierarchical two-stage vertical arbitration mechanism.

To ensure a comprehensive evaluation environment, a double-faceted simulation process was utilized to expose all designs to several traffic patterns, including request/reply memory traffic. Phase 1 of the simulation used a stand-alone, cycle-accurate NoC simulator running synthetic workloads, while Phase 2 used a hybrid NoC/cache simulator running a variety of commercial and scientific workloads within the context of a multi-bank NUCA L2 cache in an 8-CPU CMP environment. In both cases, the proposed DimDe design was demonstrated to offer average latency and throughput improvements of more than 20% over the other 3D architectures, while remaining within 5% of the full 3D crossbar performance. More importantly, the DimDe architecture outperforms all other designs, including the full 3D crossbar, by an average of 26% in terms of the Energy-Delay Product

(EDP). *One of the most important contributions of this work is the clear indication that arbitrarily adding vertical links in a 3D NoC router yields diminishing returns in terms of performance, while increasing control and arbitration complexity.*

This work, to the best of the author's knowledge, constitutes the first attempt to explore the design of a 3D-crossbar-style NoC for upcoming 3D technology. It is clear from this chapter that 3D integration presents the interconnect designer with several new challenges.

Chapter 10
Digest of Additional NoC MACRO-Architectural Research

This chapter serves as a collective summary of additional research conducted by the author in the NoC macro-architectural domain. The topics presented here were not allocated individual chapters in the volume, because they were not deemed integral to the core themes under investigation. However, the research described in this chapter is still of *significant peripheral value* to the macro-architectural pillar explored in the second part of this volume. The salient features of this chapter's work *substantially complement* the overall underlying concepts.

Specifically, *three* distinct research topics will be presented in this chapter:

(1) A distributed Multiple Entry Point (MEP) network interface architecture and its associated topological intricacies [42].
(2) A dynamic priority-based Fast Path (FP) architecture, which expedites flit traversal along frequently-used paths in the on-chip network [46].
(3) An in-depth exploration of the effects of data compression in NoCs within the context of Non-Uniform Cache Architectures (NUCA) in large-scale Chip Multiprocessors (CMP) [47].

10.1 A Distributed Multi-Point Network Interface for Low-Latency, Deadlock-Free On-Chip Interconnects [42]

In this work [42], the issues of maximizing system performance and reliability are tackled from the perspective of the network topology itself. A Multiple Entry Point topology is proposed, which provides both low-latency and inherent deadlock freedom without relying on the architecture of the routers or any proprietary and complex algorithms. In fact, the proposed topology is algorithm agnostic, and can afford its benefits to any algorithm chosen. Hence, the proposed MEP implementation is not a new router design; it is a new network topology, which can be applied to existing router designs with minimal modifications to further improve the on-chip interconnect performance. Unlike other topologies adopted in NoC architectures, which employ a *Node:Router*

C. Nicopoulos et al., *Network-on-Chip Architectures: A Holistic Design Exploration*, Lecture Notes in Electrical Engineering 45,
DOI 10.1007/978-90-481-3031-3_10, © Springer Science+Business Media B.V. 2009

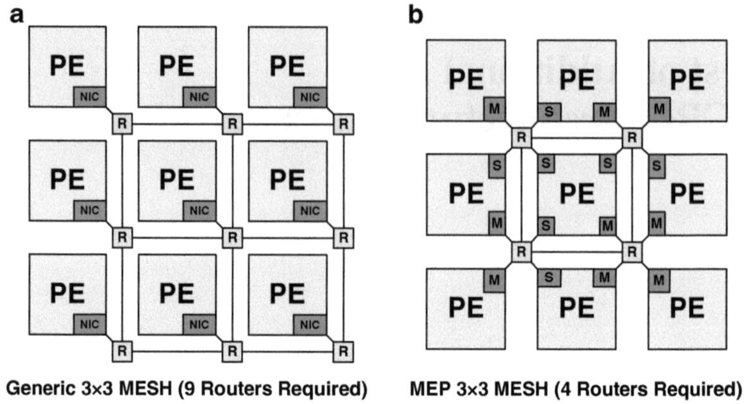

Fig. 10.1 Reduction in the number of routers when using the MEP topology (**a**) Generic 3×3 MESH (9 routers required) (**b**) MEP 3×3 MESH (4 routers required)

mapping of either *1:1* (MESH or k-ary n-cube) or *N:1* (clustered topology, where several nodes are connected to a single router), the proposed MEP topology uses *N:M* mapping. This significantly reduces the overall message latency. It also works as a framework for deadlock-free interconnection networks. Furthermore, the fact that each router connects to more than one Processing Element (PE) eliminates the need for edge routers in MESH network topologies, as illustrated in Fig. 10.1.

In an $n \times n$ network, both TORUS and MESH topologies require $n \times n$ routers. In the proposed architecture, the TORUS topology still requires $n \times n$ routers. However, in the MESH topology $(n\text{-}1) \times (n\text{-}1)$ routers are required instead, as shown in Fig. 10.1. This is a direct result of the advantageous *N:M* mapping, which allows more than one PE to be connected to a single router (Fig. 10.1b). Hence, the edge routers in a MESH topology are no longer required with the MEP mapping. On the contrary, a generic MESH requires one router per PE (Fig. 10.1a). The proposed architecture, therefore, requires *2n-1* fewer routers in an $n \times n$ MESH topology. The reduction in number of routers amounts to considerable area and energy savings, with no sacrifice in performance.

For the connection among multiple PE-nodes and routers, the generic Network Interface Controller (NIC) is divided into four smaller sub-NICs which are interconnected within a PE-node. Hence, messages in a router can be forwarded to another router via these additional paths, reducing network latency and alleviating contention. Hence, the fundamental differentiator between the MEP topology and generic implementations is the existence of two, instead of one, types of network hops in MEP. While traditional topologies transfer messages through inter-router hops, MEP employs both inter-router hops (called external hops) and sub-NIC-to-sub-NIC hops (called internal hops). To minimize the area, power, and wiring overhead, the internal hops are facilitated through the use of the lean and lightweight dTDMA bus of Chapter 7.

The MEP architecture/topology affords another inherent advantage to the on-chip network: it provides deadlock-freedom with minimal requirements from the routing algorithm. Since network behavior is not always predictable, adaptive routing algorithms are often preferred over deterministic ones in order to distribute traffic as evenly as possible over the whole network, thus avoiding traffic hot-spots and excessive congestion. But adaptive routing algorithms usually suffer from deadlocks and, hence, additional efforts to provide either deadlock recovery or avoidance are necessary. Even deterministic routing algorithms suffer from deadlock, depending on the topology. For example, the X–Y routing algorithm cannot avoid deadlock in a TORUS topology and specialized virtual channels are needed to overcome this problem. For deadlock recovery, VCs are specified such that cyclic dependency cannot occur [71]. However, such solutions rely entirely on complex routing algorithms which make the router implementation more complicated, and, consequently, area and power hungry. Software-based deadlock recovery schemes [192] require a software layer to temporarily absorb deadlocked messages out of the network. Providing a software layer may not be feasible in NoCs due to the increased implementation complexity. Instead, the proposed architecture offers deadlock freedom with only minimal requirements from the routing algorithm. This enables the routing algorithms to concentrate on traffic distribution rather than deadlock freedom, and it provides much more flexibility in algorithm design. Also, as the MEP implementation can adopt most on-chip router designs currently available (with only minimal modifications), one can expect greater latency savings when MEP is coupled with advanced router designs.

The proposed topology and NIC/router architectures were evaluated through cycle-accurate simulation and hardware synthesis implementations, and compared to those of generic topologies and architectures, in terms of latency and power consumption. Simulation results show that the proposed MEP topology and accompanying architecture can significantly reduce overall latency in both deterministic (X–Y) and minimal adaptive routing, especially at higher message injection rates. Specifically, the average latency and Energy-Delay-Product (EDP) are reduced by approximately 20% and 50%, respectively, over existing implementations. The removal of restrictions on the use of Virtual Channels for deadlock-free routing in the MEP topology model enables more efficient and versatile use of VCs, resulting in substantially reduced network contention. Finally, the additional links that connect the sub-NICs in the MEP implementation substantially improve the fault-tolerance of the communication fabric by providing redundancy in the interconnection links.

10.2 Design of a Dynamic Priority-Based Fast Path Architecture for On-Chip Interconnects [46]

The work in this project [46] revolves around a modified router architecture that accelerates flits in frequently used paths. The said architecture dynamically establishes/releases frequently used paths, based on dynamic traffic behavior, to ensure balanced utilization of valuable resources while improving overall performance. Each router maintains flit transfer statistics for each path (input–output pair) within the router for a fixed transfer interval; if some paths dominate the overall transfer, then those paths will switch to a Fast Path (FP) Mode and will be prioritized during Switch Allocation. Additionally, a pipeline bypassing technique is proposed to shorten the router pipeline along a fast path. Using this technique, the typical 2-stage router pipeline can be reduced to a single-stage pipeline.

In a typical NoC, packets tend to contend for highly utilized links; this results in network blocking and, consequently, deterioration in network performance. One way to remedy this problem is adopting clever load-balancing techniques, such as adaptive routing algorithms that dynamically route packets, based on network congestion, to reduce network blocking. Instead, the proposed Fast Path approach identifies frequently communicating patterns and then prioritizes packets that match such patterns. Hence, recurrent packets in the frequently used paths move faster, thereby reducing the probability of chained blocking. Despite the possible increase of blocking delay in low priority packets in the non-frequent paths, overall network performance can be significantly improved using this approach. Based on this premise, the proposed Fast Path (FP) architecture dynamically detects frequent communication patterns in the network and adjusts priority during Switch Allocation. The FP architecture also enables the "fast" flits to bypass the SA pipeline stage by pre-computing switch allocation one cycle before the flit arrives. Thus, a "fast" flit can directly enter the crossbar, achieving a single-stage router pipeline.

A high-level overview of the FP architecture is shown in Fig. 10.2. To determine the frequent communication patterns, the router needs to collect statistics of intra-router transfer patterns. Statistics-gathering is facilitated by a Path Frequency Analyzer (PFA) module, which consists of a set of simple counters that measure the number of actual transfers in every possible path within a router. The PFA unit initially gathers the statistics for a fixed interval. It then determines which paths are used frequently by comparing the number of transfers of each path against a threshold value.

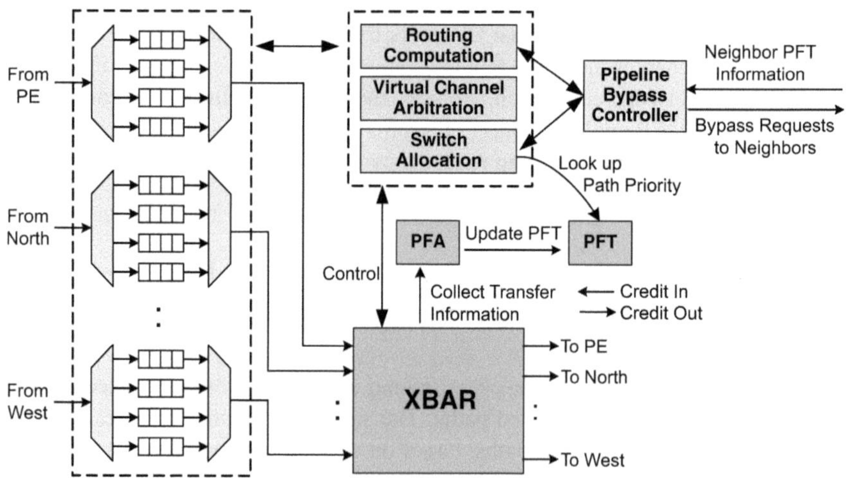

Fig. 10.2 Overview of the Fast Path architecture

If a path has transferred more flits than the threshold, it is marked as a Fast Path. The PFA updates the Path Frequency Table (PFT) accordingly, indicating which paths are designated as Fast Paths. Once a path is set as FP, a corresponding virtual channel along the FP is reserved for the "fast" flits; normal flits use the remaining VCs. The switch arbiters refer to the PFT during switch arbitration and give priority to the Fast Path VCs to expedite "fast" flit traversal. As previously mentioned, normal flits will inevitably be blocked if they are contending with "fast" flits. To avoid potential starvation of normal flits, a starvation threshold is set, such that a normal flit blocked for more than this threshold will be given the same priority as a "fast" flit. The router pipeline along the FP can be further reduced by sending the switch allocation request of a "fast" flit to the next node before the flit actually enters the node. When it does enter the node, it can immediately proceed to the crossbar stage, thus achieving a single-stage router pipeline. This operation is handled by the Pipeline Bypass Controller in Fig. 10.2.

Based on detailed hardware implementation and synthesis in 90 nm technology, the area and power overhead incurred by the proposed architecture is minimal (about 1.3% and 2.2%, respectively). In fact, the overall router power consumption decreases slightly (by around 2.5%), because the FP architecture guides flits to their destinations faster, reducing power-consuming blocking and queuing delays. In terms of sheer performance, simulation results with scientific and commercial workload traces indicate that the proposed architecture can lower network latency up to 30%, compared to a generic NoC router architecture.

10.3 Exploring the Effects of Data Compression in NoC Architectures [47]

While most prior studies in NoCs have exploited the router micro-architecture design for optimizing the NoC performance and power envelope, the essence of this work revolves around the idea of employing *data compression* [193–196] for additional gain in performance and power. The effect of compression on the interconnection network of NUCA-based CMPs is currently not clear. Toward that extent, this project analyzes the impact of compression on the underlying NoC. The ramifications on network performance, router complexity, buffer resources, and power consumption are investigated in-depth by conducting a series of detailed sensitivity experiments involving the major architectural components of a typical NoC router. Intuitively, data compression would reduce network load, and, in effect, lower average network latency and power consumption. The motivation of this work is to quantify this gain for critical design decisions. In this context, the author explores two data compression techniques for NUCA-based CMP systems, as illustrated abstractly in Fig. 10.3. The two techniques explore combinations of *storage* and *communication* compression.

The first scheme, known as *Cache Compression* (henceforth abbreviated to CC) compresses data after the L1 cache, prior to sending it to a shared L2 cache bank

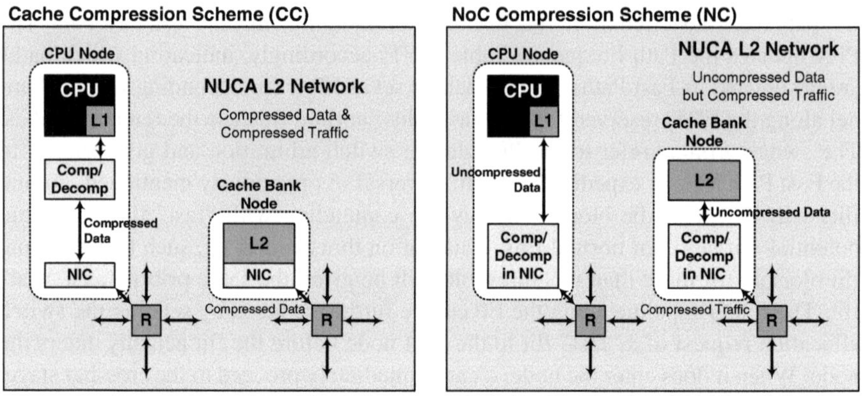

Fig. 10.3 Cache Compression (CC) vs. NoC Compression (NC)

(i.e., compression benefits both *storage* and network *communication*). Compressing data in the cache is a technique aimed at improving memory and, ultimately, system performance by squeezing more information into a fixed-size cache structure.

The second scheme, called *Compression in the NIC* (abbreviated to NC) employs compression at the Network Interface Controller, prior to injection in the network. NC implies the use of *uncompressed storage* data in the cache banks, with only the network *communication* traffic being compressed. A simple compressor/de-compressor module can be integrated in the NIC to facilitate this technique. The NC scheme has one important advantage that prompted us to include it in this work: it does not require any modification in the L2 cache banks. This attribute affords NC plug-and-play capability, which is crucial in modern System-on-Chip implementations that comprise several Intellectual Property (IP) blocks from various vendors. In many cases, these IP blocks may not be modified (known as *hard* IP cores); CC would, therefore, not be applicable in such environments. Moreover, the inclusion of NC in this study aims to characterize and quantify the effect of compression *purely in the NoC* of a NUCA CMP. On the downside, NC requires a compressor/de-compressor unit in the NIC of *every node* in the network, as can be seen in Fig. 10.3. This is in contrast to the CC scheme, which requires such units only in the *CPU nodes*. It is shown in this project that the additional area and power overhead is minimal, since the compressor/de-compressor unit is very lightweight. However, performance is also affected, because the NC mechanism involves compression and decompression operations for both read and write requests, while CC involves only one of the two operations for a given memory request (decompression for reads, compression for writes). To mitigate the performance overhead of decompression, overlap the decompression phase with network traversal. This mechanism hides some of the decompression latency overhead by commencing progressive decompression as soon as the header flit of a packet reaches its destination NIC.

To facilitate a complete system evaluation – including processor, memory, and NoC systems – an integrated NoC/cache simulator for CMP architectures was developed. A diverse set of benchmarks was used, ranging from scientific applications to commercial workloads, for in-depth performance analysis. The study starts by a co-evaluation of the NoC and processor/memory system to understand the sensitivity of the NoC design parameters on overall system performance. Then the design details of the two compression schemes (CC and NC) and their performance analyses are conducted. Overall, the two major contributions of this work are the following:

(1) To the best of the author's knowledge, this is the first work to comprehensively characterize and quantify in detail the effect of data compression on the on-chip network of a NUCA-based CMP. The effects of both CC and NC schemes on the network behavior are quantified in terms of average network latency, power consumption, and buffer utilization. Overall, CC results in network latency reduction of 24%, on average, with maximum savings of 45%. Similarly, power consumption is reduced by an average of 10% (26% maximum). Most importantly, though, we demonstrate that buffer space can be reduced by about 50% without any performance degradation. Given the dominance of on-chip router buffers in the area and power budgets (see Chapter 3), such notable reduction in buffer space is of paramount importance. NC provides network latency reduction of 7%, on average (15% maximum), while power consumption reduces by 8%, on average (14% maximum).
(2) A direct corollary of this analysis is the very interesting conclusion that existing NoC architectures are so elaborate and over-designed that they do not adequately match the speed of the processing cores. This grave mismatch in speed proves detrimental to overall performance. Instead, lean and lightweight on-chip routers are preferred in terms of operating frequency and they can over-compensate for any lack of sophistication. In fact, the study shows that elaborate attributes such as large number of virtual channels, which are considered a mainstay in modern NoC architectures, offer no tangible improvements in cache performance.

Chapter 11
Conclusions and Future Work

The continuing technology scale reduction into the deep sub-micron era has magnified the delay mismatch between gates and global wires. Wiring will significantly affect design decisions in the forthcoming billion-transistor chips, whether these are complex heterogeneous SoCs, or Chip Multi-Processors. Networks-on-Chip have surfaced as a possible solution to escalating wiring delays in future multi-core chips.

The design of efficient on-chip networks is impeded by inherently conflicting requirements: the NoC is expected to provide *ultra-low latencies*, while occupying as little *silicon real-estate*, and consuming as little *energy*, as possible. These three design strands engage in an elaborate tug-of-war, requiring extensive exploration to reach a delicate balance between all three. This intricate interplay is compounded even further by *reliability* and *variability* artifacts, which are emerging ominously as technology feature sizes dwindle. Consequently, the author presented a ***holistic approach to designing NoCs***. To fully capture the complexity underlying the interconnect architecture, the design process in this volume was guided by a quintet of fundamental design drivers: (1) **performance**, (2) **silicon area consumption**, (3) **power/energy efficiency**, (4) **reliability**, and (5) **variability**. The overall design exploration was divided into two threads: (a) **MICRO-Architectural Innovations** within the major NoC components, and (b) **MACRO-Architectural Solutions** at the system level.

These two threads formed the core guiding themes of the entire volume and led to the precipitation of a clearly defined two-part thematic structure: the first part – consisting of Chapters 3 through 6 – tackled several MICRO-architectural issues pertaining to the on-chip interconnect. More specifically, Chapter 3 introduced a new centralized buffer architecture, which dynamically allocates virtual channels and buffer slots in real-time, depending on traffic conditions. Chapter 4 presented a novel decoupled router architecture that decomposes traffic flow into two distinct and independent modules, leading to smaller and more lightweight components (including two smaller crossbars instead of a large monolithic one). The decomposition of the router into two separate micro-entities allows for graceful degradation in the presence of permanent failures. Chapter 5 delved into the world of fault-tolerance and reliability through a very detailed analysis of both transient and permanent failures within the router microarchitecture and the inter-router links. Finally, Chapter 6 constituted the first attempt

C. Nicopoulos et al., *Network-on-Chip Architectures: A Holistic Design Exploration*, Lecture Notes in Electrical Engineering, 45
DOI 10.1007/978-90-481-3031-3_11, © Springer Science+Business Media B.V. 2009

to characterize and tackle the emerging threat of Process Variation within the context of the on-chip network.

The second part of the volume – Chapters 7 through 10 – shifted attention to a higher level of abstraction, the MACRO-architecture of the NoC. Viewing the system as a whole, this section attempted to optimize the network based on its interactions with the remaining elements of the multi-core implementation. Chapter 7 experimented with the idea of hybridizing the NoC with a newly developed bus structure to extract benefits from both architectures simultaneously. Chapter 8 extended this hybridization notion to the third dimension by demonstrating that combining on-chip networks, vertical buses, and 3D architectures can be a promising option for designing large L2 cache memories for chip multiprocessors. Chapter 9 continued the author's exploration of 3D chip integration. Several interconnect architectures were evaluated by assessing their applicability and viability in a 3D setting. This evaluation culminated with the development of a new dimensionally decomposed router architecture that incorporates a true physical 3D crossbar extending through all the layers of a 3D chip. Finally, Chapter 10 presented a digest of additional research conducted by the author in the field of NoCs. This work was deemed to be of peripheral significance to the rest of the volume, but still substantial and complementary enough to be mentioned.

Overall, the work presented in this volume has significantly advanced the state-of-the-art in NoC design. Through a thorough analysis of both micro- and macro-architectural issues, the research presented here has addressed a multitude of pressing concerns spanning a wide spectrum of design topics. Improvements in all five "golden" metrics (performance, area, power, reliability, and variability) were notable and, consequently, the goal of a *holistic* design approach – set out at the beginning of the volume – has, indeed, been met. What the reader should extract from this work is the imperative need for a multi-pronged design approach, which encompasses all of these critical metrics. Designing for performance alone is no longer viable in the modern deep sub-micron epoch. Advancements in technology have spurred not only new opportunities, but also new challenges that cannot be ignored.

The design exploration in this dissertation has unveiled numerous new topics for future research in the field of NoCs. The relentless onslaught toward many-core designs requires a very sophisticated exploration of the interconnection design space. In particular, it would be extremely useful to conduct a detailed analysis/comparison between all existing and potentially new interconnect types: rings, hierarchical rings, buses, segmented buses, repeater-based buses, crossbars, mesh-based NoCs, and hybrid schemes which combine two or more of the aforementioned. Taking into consideration a diverse set of evaluation metrics, such as performance, energy, area, reliability, re-configurability, etc., one can identify efficient tradeoffs and cross-over points where one scheme becomes the best choice over the others, as a function of the number of processing cores and the underlying bandwidth requirements. Given that the interconnect is rapidly gaining a prominent role in the now prevalent multi-core design paradigm, clearly identifying the most efficient communication scheme will be crucial in meeting performance, area and power budgets in future systems. Therefore, what the author proposes as future work is the development of a very detailed framework that identifies the best interconnection type based on available resources and performance requirements. This framework can be integrated into re-configurable architectures to extend the available flexibility down to the interconnect type.

The concept of flexible interconnects by itself is certainly worth exploring. The interconnection backbone can be envisioned as a reconfigurable structure with real-time *"shape-shifting"* capabilities. The links, and even some router parameters, can adapt to prevailing network traffic conditions to enable – for instance – long and narrow express paths, or short and wider links, accordingly. The trigger for the real-time changes can be provided by network monitors on a "service" layer sitting above the main layer in a 3D environment. Such an interconnect monitor would reconfigure the network periodically to account for diverse traffic patterns in the NoC at various times.

In addition to advances in the interconnect architecture itself, the author predicts that the on-chip communication fabric will start to undertake a much more prominent role in the overall system. Given the inarguably enlarged interconnect in the complex many-core SoCs of the near future, it is reasonable to expect activities within the interconnection network that transcend the mere communication of data. This reasoning stems from the fact that data will continuously flow through the network in such communication-centric environments. Hence, some data *processing* may very well be offloaded to the network to hide some of the communication latency. The on-chip network will no longer only *transfer* data; it will also *process* it. The NoC has already been demonstrated to improve cache coherence performance in large-scale CMPs through the augmentation of the routers with additional hardware to handle in-flight, in-network cache coherence processing. Similarly, the routers could be fitted with *monitoring stations* to perform *diagnostic checks* or to detect abnormalities and, subsequently, trigger some sort of reaction mechanism. The bottom line is that the NoC routers can be upgraded with additional hardware or software modules to enable the on-chip network to extract useful information as data streams through it. By employing such techniques, the status of the NoC within the system will undoubtedly be elevated to mission-critical.

Furthermore, there are several emerging technologies that appear to be promising in terms of interconnect applicability. Carbon Nanotubes (CNT), for example, are prime candidates for future on-chip interconnection architectures. The possibility of replacing the copper interconnects with CNTs should be investigated in light of the latter's increased current density and ballistic conduction. Crosstalk and electromigration are two effects that mar existing copper wires; CNTs could possibly provide a solution to these issues, provided certain manufacturing and engineering obstacles are surpassed.

In general, the world of on-chip interconnects is brimming with countless opportunities and challenges, which will keep designers and architects busy for years to come. Clearly, an efficient interconnection backbone can open the doors to infinite possibilities in the many-core realm that has been bestowed upon us.

References

1. L. Hammond, B. A. Nayfeh, and K. Olukotun, "A single-chip multiprocessor," *In IEEE Computer*, vol. 30, pp. 79–85, 1997
2. K. Olukotun, B. Nayfeh, L. Hammond, K. Wilson, and K. Y. Chang, "The Case for a Single-Chip Multiprocessor," *in Proceedings of the 7th International Symposium on Architectural Support for Programming Languages and Operating Systems (ASPLOS)* 1996.
3. J. A. Kahle, M. N. Day, H. P. Hofstee, C. R. Johns, T. R. Maeurer, and D. Shippy, "Introduction to the Cell Multiprocessor," *IBM Journal of Research and Development*, vol. 49, pp. 589–604, 2005.
4. P. Kongetira, K. Aingaran, and K. Olukotun, "Niagara: a 32-way multithreaded Sparc processor," *in IEEE Micro*, vol. 25, pp. 21–29, 2005.
5. Intel Corporation, "Intel Develops Tera-Scale Research Chips," *http://www.intel.com/press-room/ archive/releases/20060926corp_b.htm*, September 26, 2006.
6. S. Heo and K. Asanovic, "Replacing global wires with an on-chip network: a power analysis," *in Proceedings of the 2005 International Symposium on Low Power Electronics and Design (ISLPED)*, pp. 369–374, 2005.
7. G. Smith, "Platform based design: Does it answer the entire SoC challenge?," *in Proceedings of the 41st Design Automation Conference (DAC)*, pp. 407–407, 2004.
8. International Technology Roadmap for Semiconductors (ITRS), *http://www.itrs.net/*, 2005 Edition.
9. J. M. Rabaey, A. Chandrakasan, and B. Nikolic, *Digital Integrated Circuits: A Design Perspective:* Prentice Hall, 2002.
10. R. Ho, K. W. Mai, and M. A. Horowitz, "The future of wires," *in Proceedings of the IEEE*, vol. 89, pp. 490–504, 2001.
11. P. Rickert, "Problems or opportunities? Beyond the 90nm frontier," ICCAD Keynote Address, 2004.
12. Krewell, "Multicore Showdown," *Microprocessor Report*, vol. 19, pp. 41–45, 2005.
13. L. Benini and G. D. Micheli, "Networks on Chips: A New SoC Paradigm," *IEEE Computer*, vol. 35, pp. 70–78, 2002.
14. W. J. Dally and B. Towles, "Route Packets, Not Wires: On-Chip Interconnection Networks," *in Proceedings of the Design Automation Conference (DAC)*, 2001.
15. T. D. Richardson, C. Nicopoulos, D. Park, V. Narayanan, X. Yuan, C. Das, and V. Degalahal, "A Hybrid SoC Interconnect with Dynamic TDMA-Based Transaction-Less Buses and On-Chip Networks," *in Proceedings of the International Conference on VLSI Design*, pp. 657–664, 2006.
16. A. Hemani, A. Jantsch, S. Kumar, A. Postula, J. Oberg, M. Millberg, and D. Lindqvist, "Network on chip: An architecture for billion transistor era," *in Proceedings of the IEEE NorChip Conference*, 2000.
17. P. Guerrier and A. Greiner, "A generic architecture for on-chip packet-switched interconnections," *in Proceedings of the Design, Automation and Test in Europe Conference and Exhibition (DATE)*, pp. 250–256, 2000.

18. S. Li, L. S. Peh, and N. K. Jha, "Dynamic voltage scaling with links for power optimization of interconnection networks," *in Proceedings of the 9th International Symposium on High-Performance Computer Architecture (HPCA)*, pp. 91–102, 2003.
19. R. Kumar, V. Zyuban, and D. M. Tullsen, "Interconnections in multi-core architectures: understanding mechanisms, overheads and scaling," *in Proceedings of the 32nd International Symposium on Computer Architecture (ISCA)*, pp. 408–419, 2005.
20. M. B. Taylor, J. Kim, J. Miller, D. Wentzlaff, F. Ghodrat, B. Greenwald, H. Hoffman, P. Johnson, L. Jae-Wook, W. Lee, A. Ma, A. Saraf, M. Seneski, N. Shnidman, V. Strumpen, M. Frank, S. Amarasinghe, and A. Agarwal, "The Raw microprocessor: a computational fabric for software circuits and general-purpose programs," *IEEE Micro*, vol. 22, pp. 25–35, 2002.
21. T. Simunic and S. Boyd, "Managing power consumption in networks on chips," *in Proceedings of the Design, Automation and Test in Europe Conference and Exhibition (DATE)*, pp. 110–116, 2002.
22. W. Hangsheng, L. S. Peh, and S. Malik, "Power-driven design of router microarchitectures in on-chip networks," *in Proceedings of the 36th Annual IEEE/ACM International Symposium on Microarchitecture (MICRO)*, pp. 105–116, 2003.
23. S. Li, L. S. Peh, and N. K. Jha, "PowerHerd: a distributed scheme for dynamically satisfying peak-power constraints in interconnection networks," *IEEE Transactions on Computer-Aided Design of Integrated Circuits and Systems*, vol. 25, pp. 92–110, 2006.
24. E. T. Ogawa, K. Jinyoung, G. S. Haase, H. C. Mogul, and J. W. McPherson, "Leakage, breakdown, and TDDB characteristics of porous low-k silica-based interconnect dielectrics," *in Proceedings of the IEEE International Reliability Physics Symposium*, pp. 166–172, 2003.
25. J. Srinivasan, S. V. Adve, P. Bose, and J. A. Rivers, "The impact of technology scaling on lifetime reliability," *in Proceedings of the International Conference on Dependable Systems and Networks (DSN)*, pp. 177–186, 2004.
26. E. Rosenbaum, P. M. Lee, R. Moazzami, P. K. Ko, and C. Hu, "Circuit reliability simulator-oxide breakdown module," *in Proceedings of the International Electron Devices Meeting*, pp. 331–334, 1989.
27. S. M. Alam, L. Gan Chee, C. V. Thompson, and D. E. Troxel, "Circuit level reliability analysis of Cu interconnects," *in Proceedings of the 5th International Symposium on Quality Electronic Design (ISQED)*, pp. 238–243, 2004.
28. A. T. Krishnan, V. Reddy, S. Chakravarthi, J. Rodriguez, S. John, and S. Krishnan, "NBTI impact on transistor and circuit: models, mechanisms and scaling effects [MOSFETs]," *In Technical Digest of the IEEE International Electron Devices Meeting*, pp. 14.5.1–14.5.4, 2003.
29. B. C. Paul, K. Kunhyuk, H. Kufluoglu, M. A. Alam, and K. Roy, "Temporal Performance Degradation under NBTI: Estimation and Design for Improved Reliability of Nanoscale Circuits," *in Proceedings of the Design, Automation and Test in Europe Conference and Exhibition (DATE)*, vol. 1, pp. 1–6, 2006.
30. S. Mahapatra, C. D. Parikh, V. R. Rao, C. R. Viswanathan, and J. Vasi, "Device scaling effects on hot-carrier induced interface and oxide-trapped charge distributions in MOSFETs," *IEEE Transactions on Electron Devices*, vol. 47, pp. 789–796, 2000.
31. N. Shahin, I. Ali, and P. Massoud, "Crosstalk analysis in nanometer technologies," *in Proceedings of the 16th ACM Great Lakes Symposium on VLSI (GLSVLSI)*, 2006.
32. H. H. Chen and D. D. Ling, "Power Supply Noise Analysis Methodology For Deep-submicron Vlsi Chip Design," *in Proceedings of the Design Automation Conference (DAC)*, pp. 638–643, 1997.
33. R. Baumann, "The impact of technology scaling on soft error rate performance and limits to the efficacy of error correction," *Digest of the International Electron Devices Meeting*, pp. 329–332, 2002.
34. S. Das, A. Chandrakasan, and R. Reif, "Timing, energy, and thermal performance of three-dimensional integrated circuits," *in Proceedings of the 14th ACM Great Lakes Symposium on VLSI (GLSVLSI)*, 2004.
35. R. Marculescu, "Networks-on-chip: the quest for on-chip fault-tolerant communication," *in Proceedings of the IEEE Computer Society Annual Symposium on VLSI (ISVLSI)*, pp. 8–12, 2003.

36. T. Dumitras, S. Kerner, and R. Marculescu, "Towards on-chip fault-tolerant communication," *in Proceedings of the Asia and South Pacific Design Automation Conference (ASP-DAC)*, pp. 225–232, 2003.
37. D. Park, C. Nicopoulos, J. Kim, N. Vijaykrishnan, and C. R. Das, "Exploring Fault-Tolerant Network-on-Chip Architectures," *in Proceedings of the International Conference on Dependable Systems and Networks (DSN)*, pp. 93–104, 2006.
38. E. Chang, B. Stine, T. Maung, R. Divecha, D. Boning, J. Chung, K. Chang, G. Ray, D. Bradbury, O. S. Nakagawa, S. Oh, and D. Bartelink, "Using a statistical metrology framework to identify systematic and random sources of die- and wafer-level ILD thickness variation in CMP processes," *in the Proceedings of the International Electron Devices Meeting*, pp. 499–502, 1995.
39. C. A. Nicopoulos, D. Park, J. Kim, N. Vijaykrishnan, M. S. Yousif, and C. R. Das, "ViChaR: A Dynamic Virtual Channel Regulator for Network-on-Chip Routers," *in Proceedings of the International Symposium on Microarchitecture (MICRO)*, pp. 333
40. J. Kim, C. Nicopoulos, D. Park, N. Vijakrishnan, M. S. Yousif, and C. R. Das, "A Gracefully Degrading and Energy-Efficient Modular Router Architecture for On-Chip Networks," *in Proceedings of the 33rd Annual International Symposium on Computer Architecture (ISCA)*, pp. 4–15, 2006.
41. J. Kim, D. Park, C. Nicopoulos, N. Vijaykrishnan, and C. R. Das, "Design and analysis of an NoC architecture from performance, reliability and energy perspective," *in Proceedings of the Symposium on Architecture for Networking and Communications Systems (ANCS)*, pp. 173 - 182, 200
42. D. Park, C. Nicopoulos, J. Kim, N. Vijaykrishnan, and C. R. Das, "A Distributed Multi-Point Network Interface for Low-Latency, Deadlock-Free On-Chip Interconnects," *in Proceedings of the 1st International Conference on Nano-Networks (Nano-Net)*, pp. 1–6, 2006.
43. F. Li, C. Nicopoulos, T. Richardson, Y. Xie, N. Vijakrishnan, and N. Kandemir, "Design and Management of 3D Chip Multiprocessors Using Network-in-Memory," *in Proceedings of the 33rd Annual International Symposium on Computer Architecture (ISCA)*, pp. 130–141, 2006.
44. J. Kim, C. A. Nicopoulos, D. Park, R. Das, X. Yuan, N. Vijaykrishnan, M. S. Yousif, and C. R. Das, "A novel dimensionally-decomposed router for on-chip communication in 3D architectures," *in Proceedings of the 34th Annual International Symposium on Computer Architecture (ISCA)*, pp. 138 - 149, 200
45. C.A. Nicopoulos, S. Srinivasan, A. Yanamandra, D. Park, N. Vijaykrishnan, C.R. Das, and M.J. Irwin, "On the Effects of Process Variation in Network-on-Chip Architectures," *accepted (to appear in print; published online in 10.2008) in the IEEE Transactions on Dependable and Secure Computing (TDSC).*
46. D. Park, R. Das, C. Nicopoulos, J. Kim, N. Vijaykrishnan, R. Iyer, and C. R. Das, "Design of a Dynamic Priority-Based Fast Path Architecture for On-Chip Interconnects," *in Proceedings of the Hot Interconnects Symposium*, 2007.
47. R. Das, A.K. Mishra, C.A. Nicopoulos, D. Park, N. Vijaykrishnan, R. Iyer, M.S. Yousif, C.R. Das, "Performance and Power Optimization through Data Compression in Network-on-Chip Architectures," *in Proceedings of the 14th International Symposium on High-Performance Computer Architecture (HPCA)*, pp. 215–225, 2008.
48. R. Mullins, A. West, and S. Moore, "Low-latency virtual-channel routers for on-chip networks," *in Proceedings of the International Symposium on Computer Architecture (ISCA)*, pp. 188–197, 2004.
49. L. S. Peh and W. J. Dally, "A delay model and speculative architecture for pipelined routers," *in Proceedings of the 7th International Symposium on High-Performance Computer Architecture (HPCA)*, pp. 255–266, 2001.
50. C. Xuning and L. S. Peh, "Leakage power modeling and optimization in interconnection networks," *in Proceedings of the International Symposium on Low Power Electronics and Design (ISLPED)*, pp. 90–95, 2003.
51. T. T. Ye, L. Benini, and G. De Micheli, "Analysis of power consumption on switch fabrics in network routers," *in Proceedings of the 39th Design Automation Conference (DAC)*, pp. 524–529, 2002.

52. G. Varatkar and R. Marculescu, "Traffic analysis for on-chip networks design of multimedia applications," *in Proceedings of the 39th Design Automation Conference (DAC)*, pp. 795–800, 2002.
53. H. Jingcao and R. Marculescu, "Application-specific buffer space allocation for networks-on-chip router design," *in Proceedings of the International Conference on Computer-Aided Design (ICCAD)*, pp. 354–361, 2004.
54. L. S. Peh and W. J. Dally, "A delay model for router microarchitectures," *IEEE Micro*, vol. 21, pp. 26–34, 2001.
55. W. J. Dally and C. L. Seitz, "The torus routing chip," *In Journal of Distributed Computing*, vol. 1(3), pp. 187–196, 1986.
56. P. Kermani and L. Kleinrock, "Virtual cut-through: a new computer communication switching technique," *Computer Networks*, vol. 3(4), pp. 267–286, 1979.
57. W. J. Dally, "Virtual-channel flow control," *in Proceedings of the 17th Annual International Symposium on Computer Architecture (ISCA)*, pp. 60–68, 1990.
58. W. J. Dally and C. L. Seitz, "Deadlock-free message routing in multiprocessor interconnection networks," *IEEE Transactions on Computers*, vol. C-36(5), pp. 547–553, 1987.
59. Y. M. Boura and C. R. Das, "Performance analysis of buffering schemes in wormhole routers," *IEEE Transactions on Computers*, vol. 46, pp. 687–694, 1997.
60. M. Rezazad and H. Sarbazi-azad, "The effect of virtual channel organization on the performance of interconnection networks," *in Proceedings of the 19th IEEE International Parallel and Distributed Processing Symposium*, 2005.
61. Y. Tamir and G. L. Frazier, "High-performance multiqueue buffers for VLSI communication switches," *in Proceedings of the 15th Annual International Symposium on Computer Architecture (ISCA)*, pp. 343–354, 1988.
62. G. L. Frazier and Y. Tamir, "The design and implementation of a multiqueue buffer for VLSI communication switches," *in Proceedings of the IEEE International Conference on Computer Design (ICCD)*, pp. 466–471, 1989.
63. J. Park, B. W. O'Krafka, S. Vassiliadis, and J. Delgado-Frias, "Design and evaluation of a DAMQ multiprocessor network with self-compacting buffers," *in Proceedings of Supercomputing*, pp. 713–722, 1994.
64. N. Ni, M. Pirvu, and L. Bhuyan, "Circular buffered switch design with wormhole routing and virtual channels," *in Proceedings of the International Conference on Computer Design (ICCD)*, pp. 466–473, 1998.
65. Y. Choi and T. M. Pinkston, "Evaluation of queue designs for true fully adaptive routers," *in Journal of Parallel and Distributed Computing*, vol. 64(5), pp. 606–616, 2004.
66. S. Konstantinidou and L. Snyder, "The Chaos router," *IEEE Transactions on Computers*, vol. 43, pp. 1386–1397, 1994.
67. M. Thottethodi, A. R. Lebeck, and S. S. Mukherjee, "BLAM: a high-performance routing algorithm for virtual cut-through networks," *in Proceedings of the International Parallel and Distributed Processing Symposium (IPDPS)*, pp. 10 pp., 2003.
68. A. V. Yakovlev, A. M. Koelmans, and L. Lavagno, "High-level modeling and design of asynchronous interface logic," *IEEE Design & Test of Computers*, vol. 12, pp. 32–40, 1995.
69. H. Jingcao and R. Marculescu, "Energy- and performance-aware mapping for regular NoC architectures," *IEEE Transactions on Computer-Aided Design of Integrated Circuits and Systems*, vol. 24, pp. 551–562, 2005.
70. W. Hangsheng, L. S. Peh, and S. Malik, "A technology-aware and energy-oriented topology exploration for on-chip networks," *in Proceedings of the Design, Automation and Test in Europe Conference and Exhibition (DATE)*, pp. 1238–1243 Vol. 2, 2005.
71. J. Duato, "A new theory of deadlock-free adaptive routing in wormhole networks," *in IEEE Transactions on Parallel and Distributed Systems*, vol. 4, pp. 1320–1331, 1993.
72. S. Arjun, W. J. Dally, A. K. Gupta, and B. Towles, "GOAL: a load-balanced adaptive routing algorithm for torus networks," *in Proceedings of the International Symposium on Computer Architecture (ISCA)*, pp. 194–205, 2003.

73. J. Kim, D. Park, T. Theocharides, N. Vijaykrishnan, and C. R. Das, "A low latency router supporting adaptivity for on-chip interconnects," *in Proceedings of the Design Automation Conference (DAC)*, pp. 559–564, 2005.
74. C. A. Moritz, Y. Donald, and A. Agarwal, "SimpleFit: a framework for analyzing design trade-offs in Raw architectures," *in IEEE Transactions on Parallel and Distributed Systems (TPDS)*, vol. 12, pp. 730–742, 2001.
75. L. Shang, L. S. Peh, A. Kumar, and N. K. Jha, "Thermal Modeling, Characterization and Management of On-Chip Networks," *in Proceedings of the International Symposium on Microarchitecture (MICRO)*, pp. 67–78, 2004.
76. P. Partha Pratim, C. Grecu, M. Jones, A. Ivanov, and R. Saleh, "Performance evaluation and design trade-offs for network-on-chip interconnect architectures," *in IEEE Transactions on Computers*, vol. 54, pp. 1025–1040, 2005.
77. A. Ivanov and G. De Micheli, "Guest Editors' Introduction: The Network-on-Chip Paradigm in Practice and Research," *in IEEE Design & Test of Computers*, vol. 22, pp. 399–403, 2005.
78. J. Kim, W. J. Dally, B. Towles, and A. K. Gupta, "Microarchitecture of a high radix router," *in Proceedings of the 32nd International Symposium on Computer Architecture (ISCA)*, pp. 420–431, 2005.
79. S. Chalasani and R. V. Boppana, "Fault-tolerance with multimodule routers," *in Proceedings of the International Symposium on High-Performance Computer Architecture (HPCA)*, pp. 201–210, 1996.
80. P. May, S. Bunchua, and D. S. Wills, "HiPER: a compact narrow channel router with hop-by-hop error correction," *IEEE Transactions on Parallel and Distributed Systems*, vol. 13, pp. 485–498, 2002.
81. Y. Choi and T. M. Pinkston, "Evaluation of crossbar architectures for deadlock recovery routers," *in Journal of Parallel and Distributed Computing*, vol. 61, pp. 49 – 78, 2001.
82. W. J. Dally and B. Towles, *Principles and practices of interconnection networks*: Morgan Kaufmann, 2003.
83. J. Duato, S. Yalamanchili, and L. Ni, "Interconnection networks: An engineering Approach.," *Los Alamitos, Calif., IEEE Computer Society*, 1997.
84. H. Jingcao and R. Marculescu, "Energy-aware mapping for tile-based NoC architectures under performance constraints," *in Proceedings of the Asia and South Pacific Design Automation Conference (ASP-DAC)*, pp. 233–239, 2003.
85. P. Barford and M. Crovella, "Generating representative Web workloads for network and server performance evaluation," *in Proceedings of the ACM SIGMETRICS Joint International Conference on Measurement and Modeling of Computer Systems*, 1998.
86. B. Caminero, F. J. Quiles, J. Duato, D. S. Love, and S. Yalamanchili, "Performance Evaluation of the Multimedia Router with MPEG-2 Video Traffic," *in Proceedings of the International Workshop on Communication, Architecture, and Applications for Network-Based Parallel Computing*, pp. 62–76, 1999.
87. D. Bertozzi, L. Benini, and G. De Micheli, "Low power error resilient encoding for on-chip data buses," *in Proc. of the Design, Automation and Test in Europe Conference (DATE)*, pp. 102–109, 2002.
88. A. Krstic, J. Yi-Min, and C. Kwang-Ting, "Pattern generation for delay testing and dynamic timing analysis considering power-supply noise effects," *IEEE Trans. on Computer-Aided Design of Integrated Circuits and Systems*, vol. 20, pp. 416–425, 2001.
89. K. L. Shepard and V. Narayanan, "Noise in deep submicron digital design," *in Proc. of the International Conference on Computer-Aided Design (ICCAD)*, pp. 524–531, 1996.
90. S. Murali, T. Theocharides, N. Vijaykrishnan, M. J. Irwin, L. Benini, and G. De Micheli, "Analysis of error recovery schemes for networks on chips," *in IEEE Design & Test of Computers*, vol. 22(5), pp. 434–442, 2005.
91. K. Constantinides, S. Plaza, J. Blome, Z. Bin, V. Bertacco, S. Mahlke, T. Austin, and M. Orshansky, "BulletProof: A Defect-Tolerant CMP Switch Architecture," *in Proceedings of the International Symposium on High-Performance Computer Architecture (HPCA)*, pp. 3–14, 2006.

92. P. Shivakumar, M. Kistler, S. W. Keckler, D. Burger, and L. Alvisi, "Modeling the effect of technology trends on the soft error rate of combinational logic," *in Proceedings of the International Conference on Dependable Systems and Networks (DSN)*, pp. 389–398, 2002.
93. G. V. Varatkar and R. Marculescu, "On-chip traffic modeling and synthesis for MPEG-2 video applications," *in IEEE Transactions on Very Large Scale Integration (VLSI) Systems*, vol. 12, pp. 108–119, 2004.
94. S. R. Sridhara and N. R. Shanbhag, "Coding for system-on-chip networks: a unified framework," *in Proceedings of the Design Automation Conference (DAC)*, 2004.
95. H. Zimmer and A. Jantsch, "A fault model notation and error-control scheme for switch-to-switch buses in a network-on-chip," *in Proceedings of the International Conference on Hardware/Software Codesign and System Synthesis (CODES+ISSS)*, pp. 188–193, 2003.
96. P. Vellanki, N. Banerjee, and K. S. Chatha, "Quality-of-service and error control techniques for network-on-chip architectures," *in Proceedings of the Great Lakes symposium on VLSI*, 2004.
97. M. Dall'Osso, G. Biccari, L. Giovannini, D. Bertozzi, and L. Benini, "xpipes: a latency insensitive parameterized network-on-chip architecture for multiprocessor SoCs," *in Proceedings of the 21st International Conference on Computer Design (ICCD)*, pp. 536–539, 2003.
98. W. J. Dally, L. R. Dennison, D. Harris, K. Kinhong, and T. Xanthopoulos, "Architecture and implementation of the reliable router," *in Proceedings of the Hot Interconnects Symposium*, pp. 197–208, 1994.
99. J. Wu, "A deterministic fault-tolerant and deadlock-free routing protocol in 2-D meshes based on odd-even turn model," *in Proceedings of the International Conference on Supercomputing (ICS)*, pp. 67–76, 2002.
100. T. Nesson and S. L. Johnsson, "ROMM routing on mesh and torus networks," *in Proceedings of the Symposium on Parallel Algorithms and Architectures (SPAA)*, 1995.
101. K. V. Anjan and T. M. Pinkston, "An efficient, fully adaptive deadlock recovery scheme: DISHA," *in Proceedings of the International Symposium on Computer Architecture (ISCA)*, pp. 201–210, 1995.
102. P. Liden, P. Dahlgren, R. Johansson, and J. Karlsson, "On latching probability of particle induced transients in combinational networks," *in Proceedings of the Symposium on Fault-Tolerant Computing (FTCS)*, pp. 340–349, 1994.
103. R. V. Boppana and S. Chalasani, "Fault-tolerant routing with non-adaptive wormhole algorithms in mesh networks," *in Proceedings of the Conference on Supercomputing*, pp. 693–702, 1994.
104. D. Brooks and M. Martonosi, "Dynamic thermal management for high-performance microprocessors," *in Proceedings of the International Symposium on High-Performance Computer Architecture (HPCA)*, pp. 171–182, 2001.
105. S. S. Mukherjee, F. Silla, P. Bannon, J. Emer, S. Lang, and D. Webb, "A comparative study of arbitration algorithms for the Alpha 21364 pipelined router," *in Proceedings of the International Conference on Architectural Support for Programming Languages and Operating Systems (ASPLOS)*, 2002.
106. F. Li and M. Kandemir, "Locality-conscious workload assignment for array-based computations in MPSOC architectures," *in Proceedings of the Design Automation Conference (DAC)*, pp. 95–100, 2005.
107. M. Mondal, X. Wu, A. Aziz, and Y. Massoud, "Reliability Analysis for On-chip Networks under RC Interconnect Delay Variation," *in Proceedings of the 1st International Conference on Nano-Networks (Nano-Net)*, 2006.
108. S. Borkar, T. Karnik, S. Narendra, J. Tschanz, A. Keshavarzi, and V. De, "Parameter variations and impact on circuits and microarchitecture," *in Proceedings of the Design Automation Conference (DAC)*, pp. 338–342, 2003.
109. C. Visweswariah, "Death, taxes and failing chips," *in Proceedings of the Design Automation Conference (DAC)*, pp. 343–347, 2003.

110. S. Nassif, "Delay variability: sources, impacts and trends," *in Digest of Technical Papers of the IEEE International Solid-State Circuits Conference (ISSCC)*, pp. 368–369, 2000.
111. A. Agarwal, D. Blaauw, and V. Zolotov, "Statistical timing analysis for intra-die process variations with spatial correlations," *in Proceedings of the International Conference on Computer-Aided Design (ICCAD)*, pp. 900–907, 2003.
112. Y. Peng, S. X. Shi, and D. Z. Pan, "Process variation aware OPC with variational lithography modeling," *in Proceedings of the 43rd Design Automation Conference (DAC)*, pp. 785–790, 2006.
113. L. Jiayong, L. Xin, and L. T. Pileggi, "STAC: statistical timing analysis with correlation," *in Proceedings of the 41st Design Automation Conference (DAC)*, pp. 343–348, 2004.
114. M. R. Guthaus, N. Venkateswaran, C. Visweswariah, and V. Zolotov, "Gate sizing using incremental parameterized statistical timing analysis," *in Proceedings of the International Conference on Computer-Aided Design (ICCAD)*, pp. 1029–1036, 2005.
115. A. Datta, S. Bhunia, S. Mukhopadhyay, N. Banerjee, and K. Roy, "Statistical modeling of pipeline delay and design of pipeline under process variation to enhance yield in sub-100nm technologies," *in Proceedings of the Design, Automation and Test in Europe Conference and Exhibition (DATE)*, pp. 926–931 Vol. 2, 2005.
116. Z. Wei and C. Yu, "New generation of predictive technology model for sub-45nm design exploration," *in Proceedings of the 7th International Symposium on Quality Electronic Design (ISQED)*, 2006.
117. R. Rajaraman, J. S. Kim, N. Vijaykrishnan, Y. Xie, and M. J. Irwin, "SEAT-LA: a soft error analysis tool for combinational logic," *in Proceedings of the 19th International Conference on VLSI Design*, 2006.
118. E. J. Kim, G. M. Link, K. H. Yum, N. Vijaykrishnan, M. Kandemir, M. J. Irwin, and C. R. Das, "A holistic approach to designing energy-efficient cluster interconnects," *in IEEE Transactions on Computers*, vol. 54, pp. 660–671, 2005.
119. K. Meng and R. Joseph, "Process variation aware cache leakage management," *in Proceedings of the 2006 International Symposium on Low Power Electronics and Design (ISLPED)*, pp. 262–267, 2006.
120. ARM Ltd. AMBA Specification Rev. 2.0, *http://www.arm.com/*, 1999.
121. ARM Ltd. AMBA AXI Protocol v1.0 Specification, *http://www.arm.com/*, 2004.
122. IBM Corp. 32-bit Processor Local Bus Architecture Specifications v2.9, *http://www.ibm.com/*, 2001.
123. STMicroelectronics STBus Communication System, *http://www.st.com/*, 2003.
124. L. Ruibing and K. Cheng-Koh, "SAMBA-bus: A high performance bus architecture for system-on-chips," *in Proceedings of the International Conference on Computer-Aided Design (ICCAD)*, pp. 8–12, 2003.
125. J. Y. Chen, W. B. Jone, J. S. Wang, H. I. Lu, and T. F. Chen, "Segmented bus design for low-power systems," *in IEEE Transactions on Very Large Scale Integration (VLSI) Systems*, vol. 7, pp. 25–29, 1999.
126. K. Lahiri, A. Raghunathan, and G. Lakshminarayana, "LOTTERYBUS: a new high-performance communication architecture for system-on-chip designs," *in Proceedings of the Design Automation Conference (DAC)*, pp. 15–20, 2001.
127. F. Poletti, D. Bertozzi, L. Benini, and A. Bogliolo, "Performance Analysis of Arbitration Policies for SoC Communication Architectures," *in Design Automation for Embedded Systems Journal*, vol. 8 (2–3), 2003.
128. Sonics Inc. SiliconBackplane III, *http://www.sonicsinc.com/*.
129. Sonics Inc. Sonics MX, *http://www.sonicsinc.com/*.
130. J. G. Niemann, M. Porrmann, and U. Ruckert, "A scalable parallel SoC architecture for network processors," *in Proceedings of the IEEE Computer Society Annual Symposium on VLSI (ISVLSI)*, pp. 311–313, 2005.
131. R. H. Bell, Jr., K. Chang Yong, L. John, and E. E. Swartzlander, Jr., "CDMA as a multiprocessor interconnect strategy," *In Proceedings of the 35th Asilomar Conference on Signals, Systems and Computers*, vol. 2, pp. 1246–1250 vol.2, 2001.

132. I. Verbauwhede and P. Schaumont, "The Happy Marriage of Architecture and Application in Next-Generation Reconfigurable Systems," *in Proceedings of the 1st Conference on Computing Frontiers*, pp. 363–376, 2004.
133. J. S. Lee and L. E. Miller, *CDMA Systems Engineering Handbook*: Artech House Publishers, 1998.
134. B. J. Choi, "Spreading Sequences," *http://www-mobile.ecs.soton.ac.uk/bjc97r/pnseq-1.1/index.html*, 3 May 2000.
135. R. C. Dixon, *Spread Spectrum Systems With Commercial Applications*, 3rd ed: John Wiley & Sons, Inc., 1994.
136. J. J. Komo and S. C. Liu, "Modified Kasami sequences for CDMA," *in Proceedings of the 22nd Southeastern Symposium on System Theory*, pp. 219–222, 1990.
137. K. Jongsun, X. Zhiwei, and M. F. Chang, "Reconfigurable memory bus systems using multi-Gbps/pin CDMA I/O transceivers," *in Proceedings of the International Symposium on Circuits and Systems (ISCAS)*, vol. 2, pp. II-33-II-36 vol.2, 2003.
138. P. M. Wexelblat, "An alternative addressing scheme for conventional CDMA fiber-optic networks allows interesting parallel processing capabilities," *in Proceedings of the International Conference on Parallel and Distributed Systems*, pp. 248–255, 1996.
139. R. Yoshimura, K. Tan Boon, T. Ogawa, S. Hatanaka, T. Matsuoka, and K. Taniguchi, "DS-CDMA wired bus with simple interconnection topology for parallel processing system LSIs," *in Proceedings of the IEEE International Solid-State Circuits Conference*, pp. 370–371, 2000.
140. K. Daewook, K. Manho, and G. E. Sobelman, "CDMA-based network-on-chip architecture," *in Proceedings of the IEEE Asia-Pacific Conference on Circuits and Systems*, vol. 1, pp. 137–140 vol.1, 2004.
141. A. Artieri, "Nomadik: An MPSoC Solution for Advanced Multimedia," *in Proceedings of the 5th International Forum on Application Specific MPSoC*, 2005.
142. A. Nightingale and J. Goodenough, "Testing for AMBA compliance," *in Proceedings of the IEEE International ASIC/SOC Conference*, pp. 301–305, 2001.
143. A. Hamann and R. Ernst, "TDMA time slot and turn optimization with evolutionary search techniques," *in Proceedings of the Design Automation and Test in Europe (DATE) Conference*, pp. 312–317 Vol. 1, 2005.
144. A. Maxiaguine, Z. Yongxin, C. Samarjit, and W. Weng-Fai, "Tuning SoC platforms for multimedia processing: identifying limits and tradeoffs," *in Proceedings of the International Conference on Hardware/Software Codesign and System Synthesis (CODES + ISSS)*, pp. 128–133, 2004.
145. D. J. Sorin, M. Plakal, A. E. Condon, M. D. Hill, M. M. K. Martin, and D. A. Wood, "Specifying and verifying a broadcast and a multicast snooping cache coherence protocol," *in IEEE Transactions on Parallel and Distributed Systems*, vol. 13, pp. 556–578, 2002.
146. ASU Nanoscale Integration and Modeling (NIMO) Group Predictive Technology Model (PTM), *http://www.eas.asu.edu/~ptm/*.
147. M. Coppola, "From Spaghetti wires to NoC," *in Proceedings of the 5th International Forum on Application Specific MPSoC*, 2005.
148. K. Lahiri and A. Raghunathan, "Power analysis of system-level on-chip communication architectures," *in Proceedings of the International Conference on Hardware/Software Codesign and System Synthesis (CODES + ISSS)*, pp. 236–241, 2004.
149. W. Hung, C. Addo-Quaye, T. Theocharides, Y. Xie, N. Vijakrishnan, and M. J. Irwin, "Thermal-aware IP virtualization and placement for networks-on-chip architecture," *in Proceedings of the IEEE International Conference on Computer Design (ICCD)*, pp. 430–437, 2004.
150. Kim Changkyu, Doug Burger, and S. W. Keckler, "An adaptive, non-uniform cache structure for wire-delay dominated on-chip caches," *in Proceedings of the 10th International Conference on Architectural Support for Programming Languages and Operating Systems (ASPLOS)*, 2002.

151. S. Das, A. Fan, K.-N. Chen, C. S. Tan, N. Checka, and R. Reif, "Technology, performance, and computer-aided design of three-dimensional integrated circuits," *in Proceedings of the International Symposium on Physical Design*, 2004.
152. P. Morrow, M. J. Kobrinsky, S. Ramanathan, C. M. Park, M. Harmes, V. Ramachandrarao, H. Park, G. Kloster, S. List, and S. Kim, "Wafer-Level 3D Interconnects Via Cu Bonding," *in Proceedings of the 21st Advanced Metallization Conference*, 2004
153. P. S. Magnusson, M. Christensson, J. Eskilson, D. Forsgren, and G. Hallberg, "Simics: A full system simulation platform," *in IEEE Computer*, vol. 35(2), pp. 50–58, 2002.
154. Standard Performance Evaluation Corporation, "SPEC OMP," *http://www.spec.org/hpg/omp2001/*, December 2005.
155. Z. Chishti, M. D. Powell, and T. N. Vijaykumar, "Distance associativity for high-performance energy-efficient non-uniform cache architectures," *in Proceedings of the 36th Annual IEEE/ACM International Symposium on Microarchitecture (MICRO)*, pp. 55–66, 2003.
156. Z. Chishti, M. D. Powell, and T. N. Vijaykumar, "Optimizing replication, communication, and capacity allocation in CMPs," *in Proceedings of the 32nd International Symposium on Computer Architecture (ISCA)*, pp. 357–368, 2005.
157. A. Zhang and K. Asanovic, "Victim replication: Maximizing capacity while hiding wire delay in tiled chip multiprocessors," *in Proceedings of the 32nd International Symposium on Computer Architecture (ISCA)*, 2005.
158. B. M. Beckmann and D. A. Wood, "Managing Wire Delay in Large Chip-Multiprocessor Caches," *in Proceedings of the 37th International Symposium on Microarchitecture (MICRO)*, pp. 319–330, 2004.
159. Jaehyuk Huh, Changkyu Kim, Hazim Shafi, Lixin Zhang, Doug Burger, and S. W. Keckler, "A NUCA substrate for flexible CMP cache sharing," *in Proceedings of the 19th Annual International Conference on Supercomputing (ICS)*, 2005.
160. Y. Deng and W. Maly, "2.5D system integration: a design driven system implementation schema," *in Proceedings of the Asia and South Pacific Design Automation Conference (ASP-DAC)*, pp. 450–455, 2004.
161. J. W. Joyner, P. Zarkesh-Ha, and J. D. Meindl, "A stochastic global net-length distribution for a three-dimensional system-on-a-chip (3D-SoC)," *in Proceedings of the 14th Annual IEEE International ASIC/SOC Conference*, pp. 147–151, 2001.
162. S.-M. Jung, J. Jang, W. Cho, J. Moon, K. Kwak, B. Choi, B. Hwang, H. Lim, J. Jeong, J. Kim, and K. Kim, "The revolutionary and truly 3-dimensional 25F/sup 2/ SRAM technology with the smallest S/sup 3/ (stacked single-crystal Si) cell, 0.16um/sup 2/, and SSTFT (atacked single crystal thin film transistor) for ultra high density SRAM," *in Digest of Technical Papers of the Symposium on VLSI Technology*, pp. 228–229, 2004.
163. Y. F. Tsai, Y. Xie, N. Vijaykrishnan, and M. J. Irwin, "Three-Dimensional Cache Design Exploration Using 3D Cacti," *in Proceedings of the International Conference on Computer Design (ICCD)*, 2005.
164. B. Black, D. W. Nelson, C. Webb, and N. Samra, "3D processing technology and its impact on iA32 microprocessors," *in Proceedings of the IEEE International Conference on Computer Design (ICCD)*, pp. 316–318, 2004.
165. M. Ieong, K. W. Guarini, V. Chan, K. Bernstein, R. Joshi, J. Kedzierski, and W. Haensch, "Three dimensional CMOS devices and integrated circuits," *in Proceedings of the IEEE Custom Integrated Circuits Conference*, pp. 207–213, 2003.
166. W. R. Davis, J. Wilson, S. Mick, J. Xu, H. Hua, C. Mineo, A. M. Sule, M. Steer, and P. D. Franzon, "Demystifying 3D ICs: the pros and cons of going vertical," *in IEEE Design & Test of Computers*, vol. 22, pp. 498–510, 2005.
167. A. Young, "Perspectives on 3D-IC Technology," Presentation at the 2nd Annual Conference on 3D Architectures for Semiconductor Integration and Packaging, June 2005.
168. K. Puttaswamy and G. H. Loh, "Implementing Caches in a 3D Technology for High Performance Processors," *in Proceedings of the International Conference on Computer Design (ICCD)*, pp. 525–532, 2005.

169. J. Cong and Z. Yan, "Thermal via planning for 3-D ICs," *in Proceedings of the International Conference on Computer-Aided Design (ICCAD)*, pp. 745–752, 2005.
170. B. Goplen and S. Sapatnekar, "Efficient thermal placement of standard cells in 3D ICs using a force directed approach," *in Proceedings of the International Conference on Computer-Aided Design (ICCAD)*, pp. 86–89, 2003.
171. B. Dang, P. Joseph, M. Bakir, T. Spencer, P. Kohl, and J. Meindl, "Wafer-level microfluidic cooling interconnects for GSI," *in Proceedings of the International Interconnect Technology Conference*, pp. 180–182, 2005.
172. W. L. Hung, G. M. Link, X. Yuan, N. Vijaykrishnan, and M. J. Irwin, "Interconnect and thermal-aware floorplanning for 3D microprocessors," *in Proceedings of the International Symposium on Quality Electronic Design (ISQED)*, pp. 98–104, 2006.
173. B. Black, M. Annavaram, N. Brekelbaum, J. DeVale, L. Jiang, G. H. Loh, D. McCaule, P. Morrow, D. W. Nelson, D. Pantuso, P. Reed, J. Rupley, S. Shankar, J. Shen, and C. Webb, "Die Stacking (3D) Microarchitecture," *in Proceedings of the International Symposium on Microarchitecture (MICRO)*, pp. 469–479, 2006.
174. J. Cong and Z. Yan, "Thermal-driven multilevel routing for 3D ICs," *in Proceedings of the Asia and South Pacific Design Automation Conference (ASP-DAC)*, vol. 1, pp. 121–126 Vol. 1, 2005.
175. A. Zeng, J. Lu, K. Rose, and R. J. Gutmann, "First-Order Performance Prediction of Cache Memory with Wafer-Level 3D Integration," *in IEEE Design and Test of Computers*, vol. 22(6), pp. 548–555, 2005.
176. C. C. Liu, I. Ganusov, M. Burtscher, and T. Sandip, "Bridging the processor-memory performance gap with 3D IC technology," *in IEEE Design & Test of Computers*, vol. 22, pp. 556–564, 2005.
177. T. Kgil, S. D'Souza, A. Saidi, N. Binkert, R. Dreslinski, T. Mudge, S. Reinhardt, and K. Flautner, "PicoServer: using 3D stacking technology to enable a compact energy efficient chip multiprocessor," *in Proceedings of the 12th International Conference on Architectural Support for Programming Languages and Operating Systems (ASPLOS)*, 2006.
178. S. Mysore, B. Agrawal, N. Srivastava, S. C. Lin, K. Banerjee, and T. Sherwood, "Introspective 3D chips," *in Proceedings of the 12th International Conference on Architectural Support for Programming Languages and Operating Systems (ASPLOS)*,
179. G. M. Link and N. Vijaykrishnan, "Thermal trends in emerging technologies," *in Proceedings of the 7th International Symposium on Quality Electronic Design (ISQED)*, pp. 8, 2006.
180. P. Shivakumar and N. P. Jouppi, "Cacti 3.0: An integrated cache timing, power and area model," Technical Report, Compaq Computer Corporation, August 2001.
181. Sun Microsystems Inc, "Sun UltraSPARC T1 Overview," *http://www.sun.com/processors/UltraSPARC-T1/*, December 2005.
182. P. R. Nuth and W. J. Dally, "The J-machine network," *in Proceedings of the International Conference on Computer Design (ICCD)*, pp. 420–423, 1992.
183. R. E. Kessler and J. L. Schwarzmeier, "Cray T3D: a new dimension for Cray Research," *in Proceedings of The IEEE Computer Society International Conference (COMPCON)*, pp. 176–182, 1993.
184. A. Chandrakasan, W. J. Bowhill, and F. Fox, *Design of High-Performance Microprocessor Circuits*: IEEE Press, 2001.
185. TPC-C Design Document, *http://www.tpc.org/tpcc/.*
186. SAP Sales and Distribution Benchmark, *http://www.sap.com/solutions/benchmark/index.epx.*
187. SPECjbb2005 Java Business Benchmark, *http://www.spec.org/jbb2005.*
188. SPECjAppServer Java Application Server Benchmark, *http://www.spec.org/jAppServer.*
189. S. C. Woo, M. Ohara, E. Torrie, J. P. Singh, and A. Gupta, "The SPLASH-2 programs: characterization and methodological considerations," *in Proceedings of the 22nd Annual International Symposium on Computer Architecture (ISCA)*, pp. 24–36, 1995.
190. W. J. Dally, "Virtual-channel flow control," *IEEE Transactions on Parallel and Distributed Systems*, vol. 3(2), pp. 194–205, 1992.

191. H. Sarbazi-Azad, M. Ould-Khaoua, and L. M. Mackenzie, "Analytical modelling of wormhole-routed k-ary n-cubes in the presence of matrix-transpose traffic," *in Journal of Parallel and Distributed Computing*, vol. 63, pp. 396–409, 2003.
192. J. M. Martinez-Rubio, P. Lopez, and J. Duato, "A cost-effective approach to deadlock handling in wormhole networks," *in IEEE Transactions on Parallel and Distributed Systems*, vol. 12, pp. 716–729, 2001.
193. A. R. Alameldeen and D. A. Wood, "Adaptive cache compression for high-performance processors," *in Proceedings of the 31st Annual International Symposium on Computer Architecture (ISCA)*, pp. 212–223, 2004.
194. A. R. Alameldeen and D. A. Wood, "Interactions Between Compression and Prefetching in Chip Multiprocessors," *in Proceedings of the 13th International Symposium on High-Performance Computer Architecture (HPCA)*, pp. 228–239, 2007.
195. R. B. Tremaine, T. B. Smith, M. Wazlowski, D. Har, M. Kwok-Ken, and S. Arramreddy, "Pinnacle: IBM MXT in a memory controller chip," *in IEEE Micro*, vol. 21(2), pp. 56–68, 2001.
196. K. Kant and R. Iyer, "Compressibility characteristics of address/data transfers in commercial workloads," *in Proceedings of the 5th Workshop on Computer Architecture Evaluation Using Commercial Workloads*, pp. 59–67, 2002.Short Biography

Short Biography

Chrysostomos A. Nicopoulos

Chrysostomos Nicopoulos is a Lecturer in the Department of Electrical and Computer Engineering at the University of Cyprus. Chrysostomos received the B.S. (Honors Program, minor in mathematics) and Ph.D. degrees in Electrical Engineering (specializing in Computer Engineering) from the Pennsylvania State University, USA, in 2003 and 2007, respectively. His Ph.D. advisor was Prof. Vijaykrishnan Narayanan. As a member of the Microsystems Design Laboratory at Penn State, Chrysostomos was actively involved in the research of packet-based Networks-on-Chip (NoC) and their implementation in multicore computer architectures.

His Ph.D. dissertation, presented in this book and titled: "Network-on-Chip Architectures: A Holistic Design Exploration," received the 2008 Outstanding Dissertation Award in the area of "New directions in logic and system design" by the European Design and Automation Association (EDAA).

From 2007 to 2008, he worked as a post-doctoral research associate in the Processor Architecture Laboratory at the Ecole Polytechnique Federale de Lausanne (EPFL), Switzerland. While at EPFL, Chrysostomos was involved in various research projects investigating 3D System-on-Chip (SoC) architectures, multicore computer architectures, Field Programmable Gate Arrays (FPGA), and Field Programmable Counter Arrays (FPCA).

He is currently the director of the multicore Computer Architecture Laboratory (multiCAL) at the University of Cyprus, conducting research in multi/many-core computer architectures, on-chip interconnection architectures, NoC router architectures for Chip Multi-Processors (CMP) and heterogeneous Multi-Processor Systems-on-Chip (MPSoC), 3D system architectures, embedded system architectures, and VLSI digital system design.

For more details visit:
http://www.eng.ucy.ac.cy/nicopoulos/
or
http://www.nicopoulos.eu/

MIX
Papier aus verantwortungsvollen Quellen
Paper from responsible sources
FSC® C105338

If you have any concerns about our products,
you can contact us on
ProductSafety@springernature.com

In case Publisher is established outside the EU,
the EU authorized representative is:
Springer Nature Customer Service Center GmbH
Europaplatz 3, 69115 Heidelberg, Germany

Printed by Libri Plureos GmbH
in Hamburg, Germany